Giancarlo Bernacchi

# TENSORS

## made easy

An informal introduction
to Maths of General Relativity



*To Sophie, Alexandre and Elena*

2018

$5^{th}$ edition – January 2018

ISBN 978-1-326-23097-5 (printed book)
ISBN 978-1-326-23104-0 (e-book)


Cover: photo by the author
Printed by www.lulu.com



Giancarlo Bernacchi
Rho - Milano - IT

giancarlobernacchi@libero.it

*Text edited by means of*
*Sun Microsystem OpenOffice Writer 3*
*(except some figures)*

# Contents

## Introduction

Tensor Analysis has a particular, though not exclusive, interest for the theory of General Relativity, in which curved spaces have a central role; so we cannot restrict ourselves to Cartesian tensors and to usual 3D space. In fact, all texts of General Relativity include some treatment of Tensor Analysis at various levels, but often it reduces to a schematic summary that is almost incomprehensible to the first approach, or else the topic is distributed along the text to be introduced piecemeal when needed, inevitably disorganized and fragmented. On the other hand, works devoted to the tensors at elementary level are virtually missing, neither is worth embarking on a preliminary study of specialized texts: even if one might overcome the mathematical difficulties, he would not be on the shortest way leading to relativity. What we ultimately need is a simple introduction, not too formal – maybe heuristic – suitable to be used as a first approach to the subject, before dealing with the study of specific texts of General Relativity. We will cite as references those articles or handouts that we found closer to this approach and to which we have in some measure referred in writing these notes.

Our goal is not to do anything better, but more easily affordable; we payed attention to the fundamentals and tried to make clear the line of reasoning, avoiding, as it often happens in such circumstances, too many implicit assumptions (which usually are not so obvious to the first reading).

As a general approach, we fully agree (and we do it with the faith of converts) on the appropriateness of the "geometric approach" to Tensor Analysis: really, it does not require more effort, but gives a strength to the concept of tensor that the old traditional approach "by components", still adopted by many texts, cannot give.

Addressing the various topics we have adopted a "top-down" strategy. We are convinced that it is always better, even didactically, face problems in their generality (not Cartesian tensors as introduction to general tensors, not flat spaces before curved ones); to shrink, to go into details, etc. there will always be time, but it would be better a later time. After all, there is a strategy better than others to open Chinese boxes: to begin with the larger one. This should somehow contain the irresistible temptation to unduly transfer in a general context results that apply only in particular instances.

We also tried to apply a little-known corollary of "Occam's razor": do not introduce restrictive assumptions before necessary (that is why, for example, we introduce a "dual switch" before the metric tensor). The desire not to appear pedestrian is a factor that increases by an order of magnitude the illegibility of mathematics books: we felt that it is better to let it go and accept the risk of being judged somewhat dull by those who presume to know the story in depth.

This book does not claim to originality, but aims to didactics. With an exception: the model (or metaphor) that represents the tensor as a piece or "tessera" of a mosaic (we'll call it "T-mosaic"). We are not aware that this metaphor, useful to write without embarrassment tensor equations and even a bit obvious, has ever been presented in a text. That might be even surprising. But the magicians never reveal their tricks, and mathematicians sometimes resemble them. If it will be appreciated, we will feel honored to have been just us to uncover the trick.

To read these notes should be enough some differential calculus (until partial

derivatives and Taylor expansion); for what concerns matrices, it's enough to know that they are tables with rows and columns (and that swapping them you can create a great confusion), or little more. Instead, we will assiduously use the Einstein sum convention for repeated indexes as a great simplification in writing tensor equations.
As Euclid said to the Pharaoh, there is not a royal road to mathematics; this does not mean that it is always compulsory to follow the most impervious pathway.
The hope of the author of these notes is they could be useful to someone as a conceptual introduction to the subject.

*G.B. September 2010*

**Preface to the edition 2015**

A number of problems is the substantial novelty of this new edition. Being a text dedicated to use for self-study, it seemed appropriate not only to propose the problems (with or without suggestions for the resolution) but also discuss and solve them in details. A choice imposed by the need to avoid interruptions of the text was then to group them in a second special section - in a sense autonomous. Inserting references to the problems in the text would have been difficult for the fact that a single issue usually relates to more subjects; it was decided to list in a separate index the main topics of each problem, to facilitate the reader the search for a specific topic.
The problems tend to go through, with completeness and sometimes with some more depth, the theoretical arguments presented in the first part. Various problems deal with tensor products, not a very attractive subject, but as basic as the rules of algebra can be. To curved spaces are dedicated some of the final problems, whose cut we wanted to make as intuitive as possible.
However, we believe that at first reading of the text one should not worry so much about solving problems (they may nevertheless be used to seek some clarification or explanation); only in a subsequent reading it would be important to face them in a systematic way to achieve that working knowledge without which the theoretical concepts are precariously "glued".
This new edition gave us the opportunity to introduce variants or additions and to correct some errors of the previous one.
If understanding means getting familiarity, we hope to have been an aid to achieve this goal despite the difficulty of the subject.

*G.B. May 2015*

## Notations and conventions

In these notes we'll use the standard notation for components of tensors, namely upper (apices) and lower (subscript) indexes in Greek letters, for example $T^{\alpha}_{\beta}$ , as usual in Relativity. The coordinates will be marked by upper indexes, such as $x^{\mu}$ , basis-vectors will be represented by $\vec{e}_{\alpha}$ , basis-covectors by $\tilde{e}^{\alpha}$ .

In a tensor formula such as $P_{\alpha}=g_{\alpha\beta}V^{\beta}$ the index α that appears once at left side member and once at right side member is a "free" index, while β that occurs twice in the right member only is a repeated or "dummy" index.

We will constantly use the Einstein sum convention, whereby *a repeated index means an implicit summation over that index.* For example, $P_{\alpha}=g_{\alpha\beta}V^{\beta}$ is for $P_{\alpha}=g_{\alpha 1}V^{1}+g_{\alpha 2}V^{2}+...+g_{\alpha n}V^{n}$. Further examples are: $A^{\alpha}B_{\alpha}=\sum_{\alpha=1}^{n}A^{\alpha}B_{\alpha}$ ,

$A^{\alpha\mu\nu}_{\gamma}B^{\kappa}_{\alpha\nu}=\sum_{\alpha=1}^{n}\sum_{\nu=1}^{n}A^{\alpha\mu\nu}_{\gamma}B^{\kappa}_{\alpha\nu}$ and $A^{\alpha}_{\alpha}=A^{1}_{1}+A^{2}_{2}+...+A^{n}_{n}$ .

We note that, due to the sum convention, the "chain rule" for partial derivatives can be simply written $\dfrac{\partial f}{\partial x^{\alpha}}=\dfrac{\partial f}{\partial x^{\mu}}\dfrac{\partial x^{\mu}}{\partial x^{\alpha}}$ and the sum over μ comes automatically.

A dummy index, unlike a free index, does not survive the summation and thus it does not appear in the result. Its name can be freely changed as long as it does not collide with other homonymous indexes in the same term.

In all equations *the dummy indexes must always be balanced up-down* and they cannot occur more than twice in each term. The free indexes appear only once in each term. In an equation both left and right members must have the same free indexes.

These conventions make it much easier writing correct relationships.

The little needed of matrix algebra will be said when necessary. We only note that the multiplication of two matrices requires us to "devise" a dummy index. Thus, for instance, the product of matrices $[A_{\alpha\beta}]$ and $[B^{\alpha\beta}]$ becomes $[A_{\alpha\mu}]\cdot[B^{\mu\beta}]$ (also for matrices we locate indexes up or down depending on the tensors they represent, without a particular meaning for the matrix itself).

The mark • will be used indiscriminately for scalar products between vectors and covectors, both heterogeneous and homogeneous, as well as for tensor inner products.

Other notations will be explained when introduced: there is some redundancy in notations and we will use them with easiness according to convenience. In fact, we had better getting familiar with all various alternative notations that can be found in the literature.

The indented paragraphs marked ▫ are "inserts" in the thread of the speech and they can, if desired, be skipped at first reading (they are usually justifications or proofs).

"Mnemo" boxes suggest simple rules to remind complex formulas.

"*Make things as simple as possible, but not simpler*"
A. Einstein

# 1 Vectors and covectors

## 1.1 Vectors

A set where we can define operations of addition between any two elements and multiplication of an element by a number (such that the result is still an element of the set) is called a *vector space* and its elements *vectors.* *
The usual vectors of physics are oriented quantities that fall under this definition; we will denote them generically by $\vec{V}$ .
In particular, we deal with the vector space formed by the set of vectors defined *at some point P.*

## 1.2 Basis-vectors and components

The maximum number $n$ of vectors independent to one another that we can put together is the *dimension* of a vector space. These $n$ vectors, chosen arbitrarily, form a ***basis of vectors*** (it doesn't matter they are unit or equal length vectors).
We denote the basis-vectors by $\vec{e}_1, \vec{e}_2, \dots \vec{e}_n$ (the generic basis-vector by $\vec{e}_\alpha$ , $\alpha = 1,2,3,\dots n$ and the basis as a whole by $\{\vec{e}_\alpha\}$ ).
Any other vector can be expressed as a *linear combination* of the $n$ basis-vectors. In other words, any vector can be written as an expansion (or decomposition) on the basis:

$$\vec{V} = \vec{e}_1 V^1 + \vec{e}_2 V^2 + \dots + \vec{e}_n V^n \qquad 1.1$$

that, using Einstein's sum convention, we write:

$$\vec{V} = \vec{e}_\alpha V^\alpha \qquad 1.2$$

The $n$ numbers $V^\alpha$ ($\alpha$ = 1, 2, ... $n$) are the ***components of the vector*** $\vec{V}$ upon the basis we have chosen ($\forall\ V^\alpha \in \mathbb{R}$).

> This way of representing the vector $\vec{V}$ is simply its "recipe": $\vec{e}_1, \vec{e}_2, \dots \vec{e}_n$ (i.e. the $\vec{e}_\alpha$ ) is the list of all ingredients, and $V^1, V^2, \dots V^n$ (i.e. the $V^\alpha$ ) are the quantities of each one.

The choice of the $n$ basis-vectors, provided not parallel, is arbitrary. Obviously, the components $V^\alpha$ of a given vector $\vec{V}$ change when the chosen basis changes (unlike the vector itself!)

* In fact, it must be defined $\vec{C} = a\vec{A} + b\vec{B}$ , where $\vec{A}, \vec{B}, \vec{C}$ are elements of the set and $a, b$ numbers $\in \mathbb{R}$. We gloss over other more obvious requests.

The expansion *eq.1.1* allows us to represent a vector as an *n*-tuple:

$$(V^1, V^2, \dots V^n)$$

provided there is a basis $\vec{e}_1, \vec{e}_2, \dots \vec{e}_n$ fixed in advance.

As an alternative to *eq.1.2* we will also write $\vec{V} \overset{comp}{\rightarrow} V^\alpha$ with the same meaning.

▪ As an example we graphically represent the vector-space of plane vectors ($n = 2$) that branch off from $P$, drawing only some vectors among the infinite ones:

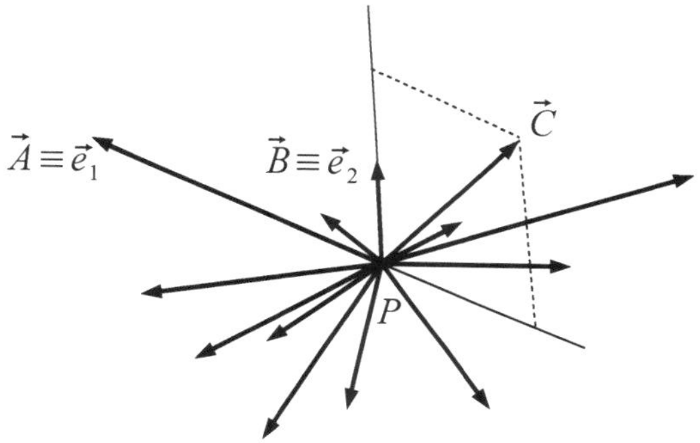


Once selected $\vec{A} \equiv \vec{e}_1$ , $\vec{B} \equiv \vec{e}_2$ as basis-vectors, a third vector $\vec{C}$ is no more independent, but it can be expressed as a linear combination of the first two; in this case:

$$\vec{C} = -\tfrac{1}{2}\vec{e}_1 + 2\vec{e}_2$$

The components of $\vec{C}$ on the established basis are thus (–½, 2).

▪ We observe that the expansion (or "recipe") of a basis-vector on its own basis (i.e. the basis made by themselves) cannot be anything but:

$$\begin{aligned} \vec{e}_1 &\rightarrow (1,0,0,\dots 0) \\ \vec{e}_2 &\rightarrow (0,1,0,\dots 0) \\ &\dots\dots\dots\dots \\ \vec{e}_n &\rightarrow (0,0,0,\dots 1) \end{aligned}$$

regardless whether they are unit basis-vector or not and their length.

## 1.3 Covectors

We define covector $\tilde{P}$ (or dual vector or "one-form") a linear scalar function* of the vector $\vec{V}$. Roughly speaking: $\tilde{P}$ applied to a vector results in a number:

$$\tilde{P}(\vec{V}) = \text{number} \in \mathbb{R} \qquad 1.3$$

$\tilde{P}$ can thus be seen as an "operator" (or "functional") that for any vector taken in input gives a number as output.
By what rule? One of the possible rules has a particular interest because it establishes a sort of reciprocity or *duality* between the vectors $\vec{V}$ and covectors $\tilde{P}$ , and it is what we aim to formalize.

▪ To begin with, we apply the covector $\tilde{P}$ to a basis-vector (instead of a generic vector). By definition, let's call the result the α-th component of $\tilde{P}$ :

$$\tilde{P}(\vec{e}_\alpha) = P_\alpha \qquad 1.4$$

We can get in this way $n$ numbers $P_\alpha$ ($\alpha = 1, 2, \dots n$), the components of the covector $\tilde{P}$ (as well as the $V^\alpha$ were the components of the vector $\vec{V}$ ).
In operational terms we can state a rule (which will be generalized later on):

> Apply the covector $\tilde{P}$ to the basis-vectors $\vec{e}_\alpha$ to get the components of the covector $\tilde{P}$ itself.

Note that this definition of component sets up a precise relationship between vectors and covectors.
Due to that definition, also a covector can be represented as an $n$-tuple

$$(P_1, P_2, \dots P_n)$$

associated with a basis $\tilde{e}^1, \tilde{e}^2, \dots \tilde{e}^n$ of covectors.
Even a covector can then be expressed as an expansion on its own basis:

$$\tilde{P} = \tilde{e}^\alpha P_\alpha \qquad (\alpha = 1, 2, \dots n) \qquad 1.5$$

using the components $P_\alpha$ as coefficients of the expansion.

---

* A function $f$ is scalar if its range is a numeric field: $f(X) =$ number.
A function $f$ is linear if $f(X+Y) = f(X) + f(Y)$ and $f(a \cdot X) = a \cdot f(X)$ or, more syntetically: $f(a \cdot X + b \cdot Y) = a \cdot f(X) + b \cdot f(Y)$

By itself, the choice of the basis of covectors is arbitrary, but at this point, having already fixed both the covector $\tilde{P}$ and (by means of *eq.1.4*) its components, the basis of covectors $\{\tilde{e}^\alpha\}$ follows, in order the last equation (*eq.1.5*) be true. In short, once used the vector basis $\{\vec{e}_\alpha\}$ to define the components of the covector, the choice of the covector basis $\{\tilde{e}^\alpha\}$ is obliged.

▪ Before giving a mathematical form to the link between the two bases, we observe that, by using the definition *eq.1.4* given above, the rule according to which $\tilde{P}$ acts on the generic vector $\vec{V}$ can be specified as:

$$\tilde{P}(\vec{V}) = \tilde{P}(\vec{e}_\alpha V^\alpha) = V^\alpha \tilde{P}(\vec{e}_\alpha) = V^\alpha P_\alpha \qquad 1.6$$

> Apply $\tilde{P}$ to $\vec{V}$ means multiply neatly in pairs the components of both and sum up.

In this way the meaning of $\tilde{P}$ is completely defined.

It is now possible to define also on the set of covectors $\tilde{P}$ operations of sum of covectors and multiplication of a covector by a number. This bestows to the set of covectors too the status of a vector space.*

We have so far defined two separate vector spaces, one for vectors, the other for covectors (both related to a point *P*), equal in dimension and in dual relationship between each other.

▪ We can now clarify how the duality relationship between the two vector spaces links the bases $\{\vec{e}_\alpha\}$ and $\{\tilde{e}^\alpha\}$.

Provided $\vec{V} = \vec{e}_\alpha V^\alpha$ and $\tilde{P} = \tilde{e}^\beta P_\beta$ we can write:

$$\tilde{P}(\vec{V}) = \tilde{e}^\beta P_\beta (V^\alpha \vec{e}_\alpha) = P_\beta V^\alpha \; \tilde{e}^\beta(\vec{e}_\alpha)$$

but, on the other hand, (*eq.1.6*):

$$\tilde{P}(\vec{V}) = P_\alpha V^\alpha$$

Both expansions are identical only if:

$$\tilde{e}^\beta(\vec{e}_\alpha) = \begin{cases} 1 & \text{for } \beta=\alpha \\ 0 & \text{otherwise} \end{cases}$$

---

* From the definition given for $\tilde{P}(\vec{V})$ it follows that: i) the application $\tilde{P}$ is linear: $\tilde{P}(a\vec{A}+b\vec{B}) = P_\alpha(aA^\alpha+bB^\alpha) = a P_\alpha A^\alpha + b P_\alpha B^\alpha = a\tilde{P}(\vec{A})+b\tilde{P}(\vec{B})$ ; ii) the set of covectors $\tilde{P}$ is also a vector space: $(a\tilde{P}+b\tilde{Q})(\vec{V}) =$ $= a\tilde{P}(\vec{V})+b\tilde{Q}(\vec{V}) = a P_\alpha V^\alpha + b Q_\alpha V^\alpha = (a P_\alpha + b Q_\alpha)V^\alpha = R_\alpha V^\alpha = \tilde{R}(\vec{V})$

After defined the *Kronecker symbol* as : $\delta^{\beta}_{\alpha} = \begin{cases} 1 & \text{for } \beta=\alpha \\ 0 & \text{otherwise} \end{cases}$ 1.7

we can write the ***duality condition***:

$$\tilde{e}^{\beta}(\vec{e}_{\alpha}) = \delta^{\beta}_{\alpha} \tag{1.8}$$

A vector space of vectors and a vector space of covectors are dual if and only if their bases are related in that way.

▪ We observe now that $\tilde{P}(\vec{V}) = V^{\alpha} P_{\alpha}$ (*eq.16*) lends itself to an alternative interpretation.
Because of its symmetry, the product $V^{\alpha} P_{\alpha}$ can be interpreted not only as $\tilde{P}(\vec{V})$ but also as $\vec{V}(\tilde{P})$, interchanging operator and operand:

$$\vec{V}(\tilde{P}) = V^{\alpha} P_{\alpha} = \text{number} \in \mathbb{R} \tag{1.9}$$

In this "reversed" interpretation a vector can be seen as a linear scalar function of a covector $\tilde{P}$. *
As a final step, if we want the duality to be complete, it would be possible to express the components $V^{\alpha}$ of a vector as the result of the application of $\vec{V}$ to the basis-covectors $\tilde{e}^{\alpha}$ (by symmetry with the definition of component of covector *eq.1.4*):

$$V^{\alpha} = \vec{V}(\tilde{e}^{\alpha}) \tag{1.10}$$

▫ It 's easy to see that this is the case, because:

$$\vec{V}(\tilde{P}) = \vec{V}(\tilde{e}^{\alpha} P_{\alpha}) = P_{\alpha} \vec{V}(\tilde{e}^{\alpha})$$

but since $\vec{V}(\tilde{P}) = P_{\alpha} V^{\alpha}$ (*eq.1.9*), equaling the right members the assumption follows.

*Eq.1.10* is the dual of *eq.1.4* and together they express a general rule:

> To get the components of a vector or covector apply the vector or covector to its *dual* basis.

## 1.4 Scalar product (heterogeneous)

The application of a covector to a vector or vice versa (the result does not change) so far denoted by notations like $\tilde{P}(\vec{V})$ or $\vec{V}(\tilde{P})$ has the meaning of a *scalar product* between the two ones. An alternative

* It's easy to verify that the application $\vec{V}$ is linear:
$\vec{V}(a\tilde{P}+b\tilde{Q}) = V^{\alpha}(a P_{\alpha}+bQ_{\alpha}) = a V^{\alpha} P_{\alpha}+b V^{\alpha} Q_{\alpha} = a \vec{V}(\tilde{P})+b\vec{V}(\tilde{Q})$

notation makes use of parentheses $\langle ... \rangle$ emphasizing the symmetry of the two operands.

The linearity of the operation is already known.
All writings:

$$\tilde{P}(\vec{V}) = \vec{V}(\tilde{P}) = \langle \tilde{P}, \vec{V} \rangle = \langle \vec{V}, \tilde{P} \rangle = V^{\alpha} P_{\alpha} \qquad 1.11$$

are equivalent and represent the heterogeneous scalar product between a vector and a covector.
By the new introduced notation the duality condition between bases *eq.1.8* is currently expressed as:

$$\langle \tilde{e}^{\beta}, \vec{e}_{\alpha} \rangle = \delta^{\beta}_{\alpha} \qquad 1.12$$

The homogeneous scalar product between vectors or covectors of the same kind requires a different definition that will be given further on.

## 1.5 The Kronecker δ

*Eq.1.7* defines the Kronecker δ: $\delta^{\beta}_{\alpha} = \begin{cases} 1 & \text{for } \beta = \alpha \\ 0 & \text{otherwise} \end{cases}$

A remarkable property of the Kronecker δ often used in practical calculations is that $\delta^{\beta}_{\alpha}$ acts as an operator that identifies the two indexes α, β turning one into the other.

For example: $\delta^{\beta}_{\alpha} V^{\alpha} = V^{\beta}$ ; $\delta^{\beta}_{\alpha} P_{\beta} = P_{\alpha}$ 1.13

Note that the summation that is implicit in the first member collapses to the single value of the second member. This happens because $\delta^{\beta}_{\alpha}$ removes from the sum all terms whose indexes α, β are different, making them equal to zero.
Roughly speaking: $\delta^{\beta}_{\alpha}$ hooks by its first index the index of the operand and changes it into its own second index: what survives is the free index (note the "balance" of the indexes in both *eq.1.13*).

▫ We prove the first of *eq.1.13*:
Multiplying $\delta^{\beta}_{\alpha} = \tilde{e}^{\beta}(\vec{e}_{\alpha})$ (*eq.1.8*) by $V^{\alpha}$ gives:
$\delta^{\beta}_{\alpha} V^{\alpha} = \tilde{e}^{\beta}(\vec{e}_{\alpha}) V^{\alpha} = \tilde{e}^{\beta}(\vec{V}) = V^{\beta}$
(the last equality of the chain is the rule *eq.1.10*).
Similarly for the other equation *eq.1.13*.

In practice, this property of Kronecker δ turns useful each time we can

make a product like $\tilde{e}^{\beta}(\vec{e}_{\alpha})$ to appear in an expression: we can replace it by $\delta_{\alpha}^{\beta}$ , that soon produces a change of the index in one of the factors, as shown in *eq.1.13*.
So far we have considered the Kronecker symbol $\delta_{\alpha}^{\beta}$ as a number; we will see later that $\delta_{\alpha}^{\beta}$ is to be considered as a component of a tensor.

## 1.6 Metaphors and models

The fact that vectors and covectors can both be represented as a double array of basis elements and components

$$\begin{matrix} \vec{e}_1 & \vec{e}_2 & \vec{e}_3 & \dots & \vec{e}_n \\ V^1 & V^2 & V^3 & \dots & V^n \end{matrix} \qquad \text{and} \qquad \begin{matrix} \tilde{e}^1 & \tilde{e}^2 & \tilde{e}^3 & \dots & \tilde{e}^n \\ P_1 & P_2 & P_3 & \dots & P_n \end{matrix}$$

suggests an useful metaphor to display graphically the formulas.

▪ Vectors and covectors are represented as interlocking cubes or "building blocks" bearing pins and holes that allow to hookup each other.

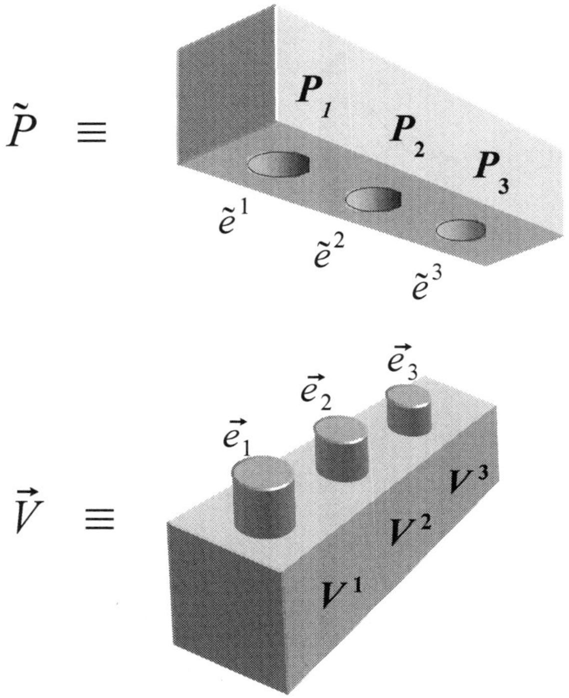

Vectors are cubes with pins upward; covectors are cubes with holes downward. We'll refer to pins and holes as "connectors".
Pins and holes are in number of $n$, the dimension of space.
Each pin represents a $\vec{e}_\alpha$ ($\alpha = 1, 2, \ldots n$); each hole is for a $\tilde{e}^\alpha$.
In correspondence with the various pins or holes we have to imagine the respective components written sideways on the body of the cube, as shown in the picture.
The example above refers to a 3D space (for $n = 4$ the cubes would have 4 pins or holes, and so on).
The $n$ pins and the $n$ holes must connect together simultaneously. Their connection emulates the heterogeneous scalar product between vectors and covectors and creates an object with no exposed connectors (a scalar). The heterogeneous scalar product

$$\left.\begin{array}{l} \tilde{e}^\alpha P_\alpha \\ \\ \vec{e}_\alpha V^\alpha \end{array}\right\} \xrightarrow[\text{product}]{\text{scalar}} \quad P_\alpha V^\alpha$$

may be indeed put in a pattern that may be interpreted as the metaphor of interlocking cubes:

$$\left.\begin{array}{l} \begin{array}{ccccccccc} \tilde{e}^1 & & \tilde{e}^2 & & \tilde{e}^3 & & & \tilde{e}^n \\ \times & + & \times & + & \times & \cdots & & \times \\ P_1 & & P_2 & & P_3 & & & P_n \end{array} \\ \\ \begin{array}{ccccccccc} \vec{e}_1 & & \vec{e}_2 & & \vec{e}_3 & & & \vec{e}_n \\ \times & + & \times & + & \times & \cdots & & \times \\ V^1 & & V^2 & & V^3 & & & V^n \end{array} \end{array}\right\} \xrightarrow[\text{product}]{\text{scalar}} \quad \begin{array}{ccccccccc} P_1 & & P_2 & & P_3 & & & P_n \\ \times & + & \times & + & \times & \cdots & & \times \\ V^1 & & V^2 & & V^3 & & & V^n \end{array}$$

as shown in the previous picture, once we imagine to have connected the two cubes.

## 1.7 The "T-mosaic" model

In the following we will adopt a more simple drawing without perspective, representing the cube as seen from the short side, like a piece of a mosaic (a "tessera") in two dimensions with no depth, so that all the pins and holes will be aligned along the line of sight and merge into one which represents them collectively. The generic name of the array ($\vec{e}_\alpha$ for pins or $\tilde{e}^\alpha$ for holes) is thought as written upon it, while the generic name of the component array ( $V^{\alpha}$ for vectors or $P_{\alpha}$ for covectors) is written in the body of the piece. This representation is the basis of the model that we call "**T-mosaic**" for its ability to be generalized to the case of tensors.

▪ So, from now on we will use the two-dimensional representation:

Vector $\vec{V} \equiv$ 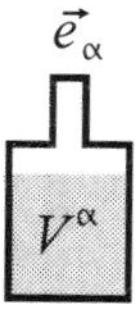


Covector $\tilde{P} \equiv$ 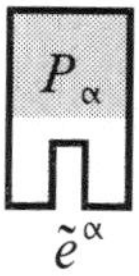


Blocks like those in the figure give a synthetic representation of the expansion by components on the given basis (the "recipe" *eq.1.2*, *eq.1.5*).

▪ The connection of the blocks, i.e. the connection of a block to another, represents the *heterogeneous scalar product:*

$$\tilde{P}(\vec{V}) = \vec{V}(\tilde{P}) = \langle \tilde{P}, \vec{V} \rangle \equiv$$

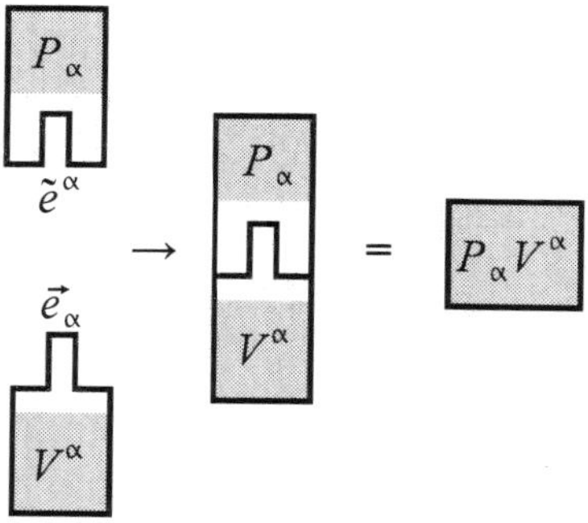


The connection is made between homonymous connectors ( = with the same index: in this example $\tilde{e}^{\alpha}$ and $\vec{e}_{\alpha}$ ). When the blocks fit together, the connectors disappear and bodies merge *multiplying to each other.*

▪ Basis-vector will be drawn as: 

and basis-covector as: 

For simplicity, nothing will be written in the body of the tessera which represents a basis-vector or basis-covector, but we have to remember that, according to the perspective representation, a series of 0 together with a single 1 are inscribed in the side of the block.
Example in 5D:

$\vec{e}_2 \equiv$

This means that, in the scalar product, all the products that enter in the sum go to zero, except one.

▪ The application of a covector to a basis-vector to get the covector components (*eq.1.4*) is rendered by the connection:

$\tilde{P}(\vec{e}_\alpha) = P_\alpha$ :

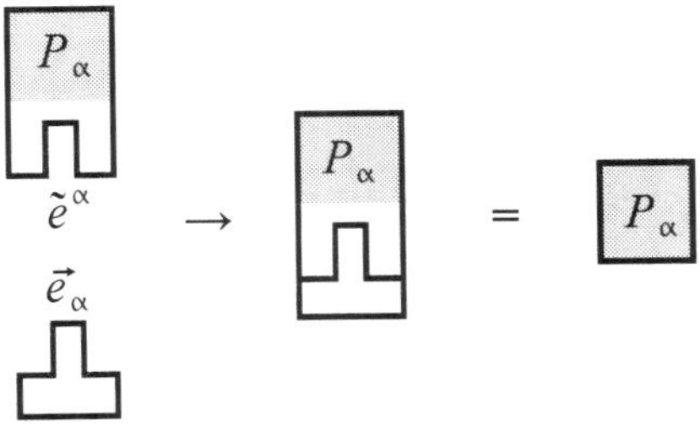


Similarly, the application of a vector to a basis-covector to get the components of the vector (*eq.1.10*) has as image the connection:

$\vec{V}(\tilde{e}^\alpha) = V^\alpha$ :

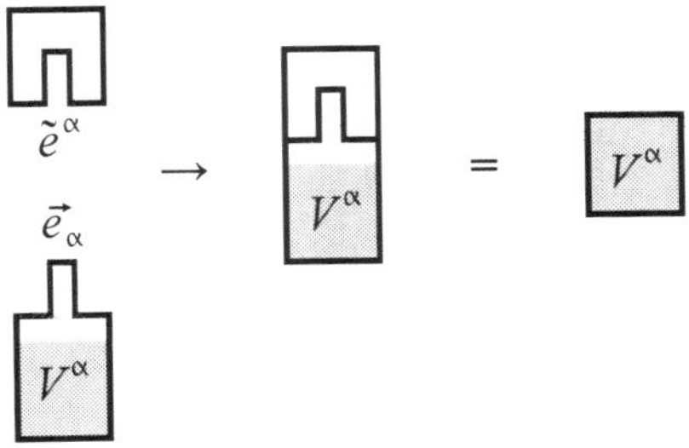

A "smooth" block, i.e. a block without free connectors, is a scalar ( = a number).

▪ Note that:

> In T-mosaic representation the connection always occurs between connectors with the same index (same name), in contrast to what happens in algebraic formulas (where it is necessary to diversify them in order to avoid the summations interfere with each other).

▪ However, it is still possible to perform blockwise the connection between different indexes, in which case we have to insert a block of Kronecker δ as a "*plug adapter*":

$$\tilde{P}(\vec{A}) = \underbrace{P_\alpha \tilde{e}^\alpha (A^\beta \vec{e}_\beta)} = P_\alpha A^\beta \underbrace{\tilde{e}^\alpha(\vec{e}_\beta)}_{\delta^\alpha_\beta} = \underbrace{P_\alpha A^\beta \delta^\alpha_\beta} = \underbrace{P_\alpha A^\alpha}$$

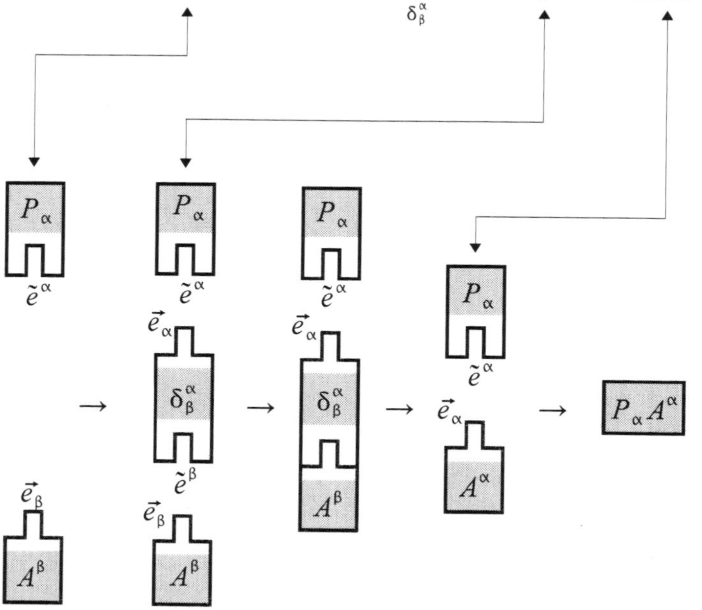


The chain of equalities above is the usual sequence of steps when using the conventional algebraic notation, which makes use of different indexes and the symbol Kronecker δ.
Note the correspondence between successive steps in algebraic formulas and blocks.

It is worth to note that in the usual T-mosaic representation the connection occurs directly between $P_\alpha \tilde{e}^\alpha$ and $A^\alpha \vec{e}_\alpha$ by means of the homonymic α-connectors, skipping the first 3 steps.

▪ The T-mosaic representation of the duality relation (*eq.1.8* or *eq.1.12*) is a significant example of the use of Kronecker δ as a "plug adapter":

$$\langle \tilde{e}^\beta, \vec{e}_\alpha \rangle \equiv$$

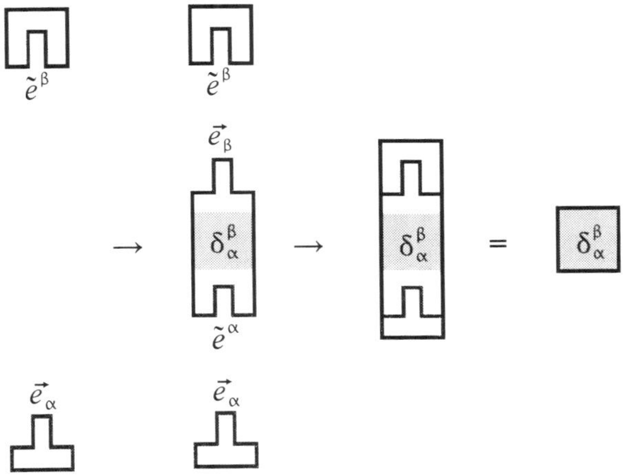


In practice, it does not matter plugging directly connectors of the same name or connectors with different index via an interposed plug adapter; the first way is easier and faster, the second allows a better correspondence with algebraic formulas.
We will see later that using the Kronecker δ as a "plug adapter" block is justified by its tensor character.

## 2 Tensors

The concept of tensor $\mathbf{T}$ is an extension of those of vector and covector.

A tensor is a linear scalar function of $h$ covectors and $k$ vectors ($h$, $k$ = 0, 1, 2, ...).

We may see $\mathbf{T}$ as an operator that takes $h$ covectors and $k$ vectors in input to give a number as a result:

$$\mathbf{T}(\vec{A}, \vec{B}, \ldots \tilde{P}, \tilde{Q}, \ldots) = \text{number} \in \mathbb{R} \qquad 2.1$$

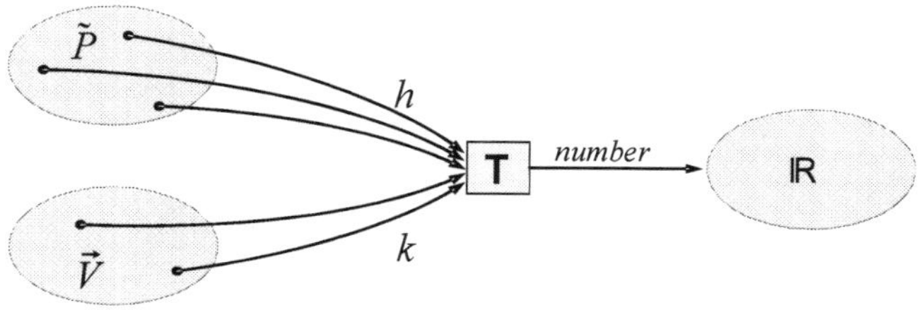


By $\binom{h}{k}$ or even $r = h+k$ we denote the rank or order of the tensor.

A tensor $\binom{0}{0}$ is a scalar; a tensor $\binom{1}{0}$ is a vector; a tensor $\binom{0}{1}$ is a covector.

*Warning*: in the notation $\mathbf{T}(\ ,\ , \ldots)$ the brackets following the symbol of tensor contain the list of arguments or objects that the tensor takes as input (*input list*). It does not simply "qualify" the tensor, but it represents an operation already performed. In fact, $\mathbf{T}(\ ,\ , \ldots)$ is the *result* of applying $\mathbf{T}$ to the list within parentheses. The "naked tensor" is simply written $\mathbf{T}$.

▪ The *components of a tensor* $\mathbf{T}$ are defined in a way similar to vectors and covectors by applying the tensor to basis-vectors and covectors. In simple words: we input into the tensor $\mathbf{T}$ the required amount of basis-vectors and basis-covectors ($h$ basis-covectors and $k$ basis-vectors); the number that comes out is the component of the tensor on the given bases. For example, the components of a $\binom{0}{2}$ tensor result by giving to the tensor the various pairs $\vec{e}_\alpha, \vec{e}_\beta$ :

$$\mathbf{T}(\vec{e}_\alpha, \vec{e}_\beta) = T_{\alpha\beta} \qquad 2.2$$

in order to obtain $n^2$ numbers marked by the double index $\alpha\,\beta$ (for instance, giving $\vec{e}_1, \vec{e}_2$ we get $T_{12}$ , and so on).

In general:

$$\mathbf{T}(\vec{e}_\alpha, \vec{e}_\beta, \ldots \tilde{e}^\mu, \tilde{e}^\nu, \ldots) = T^{\mu\nu\ldots}_{\alpha\beta\ldots} \qquad 2.3$$

▪ This equation allows us to specify the calculation rule left undetermined by *eq.2.1* (which number does come out by applying the tensor to an input list?). For example, given a tensor **S** of rank $\binom{1}{1}$ for which *eq.2.3* becomes $\mathbf{S}(\vec{e}_\alpha, \tilde{e}^\gamma) = S^\gamma_\alpha$ , we get:

$$\mathbf{S}(\vec{V}, \tilde{P}) = \mathbf{S}(\vec{e}_\alpha V^\alpha, \tilde{e}^\gamma P_\gamma) = V^\alpha P_\gamma \mathbf{S}(\vec{e}_\alpha, \tilde{e}^\gamma) = V^\alpha P_\gamma S^\gamma_\alpha \qquad 2.4$$

In general *eq.2.1* works as follows:

$$\mathbf{T}(\vec{A}, \vec{B}, \ldots \tilde{P}, \tilde{Q}, \ldots) = A^\alpha B^\beta \cdots P_\mu Q_\nu \cdots T^{\mu\nu\ldots}_{\alpha\beta\ldots} \qquad 2.5$$

and its result is a number (one for each set of values μ, ν, α, β, ...).

> An expression like $A^\alpha B^\beta P_\mu Q_\nu T^{\mu\nu}_{\alpha\beta}$ that contains only balanced dummy indexes is a scalar because no index survives in the result of the implicit multiple summation.

*Eq.2.4*, *eq.2.5* appear as an extension to tensors of *eq.1.6*, *eq.1.9* valid for vectors and covectors. To apply the tensor **T** to its input list ( , , ...) looks like the application $\vec{V}(\tilde{P})$ or $\tilde{P}(\vec{V})$ of a vector or covector to its respective argument, meaning heterogeneous scalar product. It may be seen as a sort of multiple scalar product, i.e. a sequence of scalar products between the tensor and the vectors / covectors of the list (indeed, it is a tensor "inner product" similar to the scalar product of vectors and covectors as we'll explain later on).

▪ Speaking of tensor components we have so far referenced to bases of vectors and covectors. However, it is possible to express the tensor as an expansion by components on *its own basis*. That is to say, in the case of a $\binom{0}{2}$ tensor:

$$\mathbf{T} = \tilde{e}^{\alpha\beta} T_{\alpha\beta} \qquad 2.6$$

(it's the "recipe" of the tensor, similar to that of vectors). This time the basis is a tensorial one and consists of basis-tensors $\tilde{e}^{\alpha\beta}$ with double index, in number of $n^2$.

This expression has only a formal meaning until we specify what sort of basis is that and how it is related to the basis of vectors / covectors already defined. To do this we have to define first the outer product between vectors and /or covectors.

## 2.1 Outer product between vectors and covectors

Given the vectors $\vec{A}, \vec{B}$ and the covectors $\tilde{P}, \tilde{Q}$ we define a *vector outer product* between the vectors $\vec{A}$ and $\vec{B}$:

$$\vec{A} \otimes \vec{B} \text{ such that: } \vec{A} \otimes \vec{B}(\tilde{P}, \tilde{Q}) = \vec{A}(\tilde{P})\ \vec{B}(\tilde{Q}) \qquad 2.7$$

Namely: $\vec{A} \otimes \vec{B}$ is an operator acting on a *couple of covectors* (i.e. vectors of the opposite kind) in terms of two scalar products, as stated in the right member. The result is here again a number $\in \mathbb{R}$ .

It can be immediately seen that the outer product $\otimes$ is *non-commutative*: $\vec{A} \otimes \vec{B} \neq \vec{B} \otimes \vec{A}$ (indeed, note that $\vec{B} \otimes \vec{A}$ against the same operand would give as a different result $\vec{B}(\tilde{P})\ \vec{A}(\tilde{Q})$ ).

Also note that $\vec{A} \otimes \vec{B}$ is a rank $\binom{2}{0}$ tensor because it matches the definition given for a tensor: it takes 2 covectors as input and gives a number as result.

Similarly we can define the outer products between covectors or between vectors and covectors, ranked $\binom{0}{2}$ and $\binom{1}{1}$ rispectively:

$$\tilde{P} \otimes \tilde{Q} \text{ such that: } \tilde{P} \otimes \tilde{Q}\ (\vec{A}, \vec{B}) = \tilde{P}(\vec{A})\ \tilde{Q}(\vec{B}) \quad \text{and also}$$
$$\tilde{P} \otimes \vec{A} \text{ such that: } \tilde{P} \otimes \vec{A}\ (\vec{B}, \tilde{Q}) = \tilde{P}(\vec{B})\ \vec{A}(\tilde{Q}) \quad \text{etc.}$$

▪ Starting from vectors and covectors and making outer products between them we can build tensors of gradually increasing rank.
In general, the inverse is *not* true: not all tensors can be expressed as outer product of tensors of lower rank.

▪ We can now characterize the tensor-basis in terms of outer product of basis-vectors and / or covectors. To fix on the case $\binom{0}{2}$ , it is:

$$\tilde{e}^{\mu\nu} = \tilde{e}^{\mu} \otimes \tilde{e}^{\nu} \qquad 2.8$$

▫ In fact, * from the definition of component $T_{\alpha\beta} = \mathbf{T}(\vec{e}_\alpha, \vec{e}_\beta)$ (*eq.2.2*), using the expansion $\mathbf{T} = T_{\mu\nu}\ \tilde{e}^{\mu\nu}$ (*eq.2.6*) we get:

$T_{\alpha\beta} = T_{\mu\nu}\ \tilde{e}^{\mu\nu}(\vec{e}_\alpha, \vec{e}_\beta)$ , which is true only if:

---

* The most direct demonstration, based on a comparison between the two forms $\mathbf{T} = \tilde{P} \otimes \tilde{Q} = P_\alpha \tilde{e}^\alpha \otimes Q_\beta \tilde{e}^\beta = P_\alpha Q_\beta \tilde{e}^\alpha \otimes \tilde{e}^\beta = T_{\alpha\beta} \tilde{e}^\alpha \otimes \tilde{e}^\beta$ and $\mathbf{T} = T_{\alpha\beta}\ \tilde{e}^{\alpha\beta}$ holds only in the case of tensors decomposable as tensor outer product.

$\tilde{e}^{\mu\nu}(\vec{e}_\alpha, \vec{e}_\beta) = \delta^\mu_\alpha \delta^\nu_\beta$ because in this case:

$T_{\alpha\beta} = T_{\mu\nu} \; \delta^\mu_\alpha \delta^\nu_\beta$ .

Since $\delta^\mu_\alpha = \tilde{e}^\mu(\vec{e}_\alpha)$ and $\delta^\nu_\beta = \tilde{e}^\nu(\vec{e}_\beta)$ the second-last equation becomes:

$\tilde{e}^{\mu\nu}(\vec{e}_\alpha, \vec{e}_\beta) = \tilde{e}^\mu(\vec{e}_\alpha) \; \tilde{e}^\nu(\vec{e}_\beta)$

which, by definition of $\otimes$ (*eq.2.7*), means:

$\tilde{e}^{\mu\nu} = \tilde{e}^\mu \otimes \tilde{e}^\nu$ , q.e.d.

Hence, the basis of tensors has been reduced to the basis of (co)vectors. The $\binom{0}{2}$ tensor under consideration can then be expanded on the basis of covectors:

$$\mathbf{T} = T_{\alpha\beta} \; \tilde{e}^\alpha \otimes \tilde{e}^\beta \qquad 2.9$$

It's again the "recipe" of the tensor, as well as *eq.2.6*, but this time it uses basis-vectors and basis-covectors as "ingredients".

▪ In general, a tensor can be written as a linear combination of (or as an expansion over the basis of) elementary outer products $\vec{e}_\alpha \otimes \vec{e}_\beta \otimes \ldots \tilde{e}^\mu \otimes \tilde{e}^\nu \otimes \ldots$ whose coefficients are the components.

For instance, a $\binom{3}{1}$ tensor can be expanded as:

$$\mathbf{T} = T^{\alpha\beta\gamma}_{\mu} \; \vec{e}_\alpha \otimes \vec{e}_\beta \otimes \vec{e}_\gamma \otimes \tilde{e}^\mu \qquad 2.10$$

which is usually simply written:

$$\mathbf{T} = T^{\alpha\beta\gamma}_{\mu} \; \vec{e}_\alpha \, \vec{e}_\beta \, \vec{e}_\gamma \, \tilde{e}^\mu \qquad 2.11$$

Note the "balance" of upper / lower indexes.

The symbol $\otimes$ can be normally omitted without ambiguity.

▫ In fact, products $\vec{A}\vec{B}$ or $\vec{V}\tilde{P}$ can be unambiguously interpreted as $\vec{A} \otimes \vec{B}$ or $\vec{V} \otimes \tilde{P}$ because for other products other explicit symbols are used, such as $\vec{V}(\tilde{P})$ , $\langle \vec{V}, \tilde{P} \rangle$ or $\vec{V} \bullet \vec{W}$ for scalar products, $\mathbf{T}(\ldots)$ or even $\bullet$ for tensor inner products.

▪ ***The order of the indexes*** is important and is stated by *eq.2.10* or *eq.2.11* which represent the tensor in terms of an outer product: it is understood that changing the order of the indexes means to change the order of factors in the outer product, in general non-commutative.

Thus, in general $T^{\alpha\beta\gamma} \neq T^{\beta\alpha\gamma} \neq T^{\gamma\alpha\beta} \dots$ .

To preserve the order of the indexes a notation with a double sequence for upper and lower indexes like $Y^{\alpha\gamma\mu}_{\beta\nu}$ is often enough. However, this notation is ambiguous and turns out to be improper when indexes are raised / lowered. Actually, it would be convenient to use a scanning with reserved columns like $Y\left|{}^{\alpha}_{\cdot}\right|{}^{\cdot}_{\beta}\left|{}^{\gamma}_{\cdot}\right|$ ; it aligns in a single sequence upper and lower indexes and assigns to each index a specific place wherein it may move up and down without colliding with other indexes. To avoid any ambiguity we ought to use a notation such as $Y^{\alpha\cdot\gamma\mu\cdot}_{\cdot\beta\cdot\cdot\nu}$ , where the dot $\cdot$ is used to keep busy the column, or simply $Y^{\alpha\ \gamma\mu}_{\ \beta\ \ \nu}$ replacing the dot with a blank.

## 2.2 Matrix representation of tensors

- A tensor of rank 0 (a scalar) is a number
- a tensor of rank 1, that is $\binom{0}{1}$ or $\binom{1}{0}$ , is an $n$-tuple of numbers
- a tensor of rank 2, that is $\binom{2}{0}, \binom{1}{1}, \binom{0}{2}$ , is a square matrix $n \times n$
- a tensor of rank 3 is a "cubic lattice" of $n \times n \times n$ numbers, etc...

(In each case the numbers are the components of the tensor).

A tensor of rank $r$ can be thought of as an $r$-dimensional grid of numbers, or components (a single number or $n$-array or $n \times n$ matrix, etc.); in all cases we must think the component grid as associated with an underlying "bases grid" of the same dimension.

On the contrary, not every $n$-tuple of numbers or $n \times n$ matrix, and so on, is a vectors, tensor, etc.

Only tensors of rank 1 and 2, since represented as vectors or matrices, can be treated with the usual rules of matrix calculus (the advantage to do that may be due to the fact that the inner product between tensors of these ranks is reduced to the product of matrices).

In particular, tensors $\binom{2}{0}$ , $\binom{1}{1}$ or $\binom{0}{2}$ can all be represented by matrices, although built on different bases grids. For instance, for tensors of rank $\binom{2}{0}$ the basis grid is:

$$\begin{matrix} \vec{e}_1 \otimes \vec{e}_1 & \vec{e}_1 \otimes \vec{e}_2 & \cdots & \vec{e}_1 \otimes \vec{e}_n \\ \vec{e}_2 \otimes \vec{e}_1 & \vec{e}_2 \otimes \vec{e}_2 & & \vec{e}_2 \otimes \vec{e}_n \\ \vdots & & & \vdots \\ \vec{e}_n \otimes \vec{e}_1 & \vec{e}_n \otimes \vec{e}_2 & \cdots & \vec{e}_n \otimes \vec{e}_n \end{matrix}$$

On this basis the matrix of the tensor is:

$$\mathbf{T} \equiv \begin{bmatrix} T^{11} & T^{12} & \cdots & T^{1n} \\ T^{21} & T^{22} & & T^{2n} \\ \vdots & & & \vdots \\ T^{n1} & T^{n2} & \cdots & T^{nn} \end{bmatrix} = \left[ T^{\alpha\beta} \right] \qquad 2.12$$

Similarly:

for $\binom{1}{1}$ tensor: $\mathbf{T} \equiv \left[ T^{\alpha}_{\beta} \right]$ on grid $\vec{e}_\alpha \otimes \tilde{e}^\beta$

for $\binom{0}{2}$ tensor: $\mathbf{T} \equiv \left[ T_{\alpha\beta} \right]$ on grid $\tilde{e}^\alpha \otimes \tilde{e}^\beta$

▪ The outer product between vectors and tensors has similarities with the Cartesian product (= set of pairs).
For example $\vec{V} \otimes \vec{W}$ generates a basis grid $\vec{e}_\alpha \otimes \vec{e}_\beta$ that includes all the pairs that can be formed by $\vec{e}_\alpha$ and $\vec{e}_\beta$ and a related matrix of components with all possible pairs such as $V^\alpha W^\beta \left( = T^{\alpha\beta} \right)$, all ordered by row and column.

If $\mathbf{X}$ is a rank $\binom{2}{0}$ tensor, the outer product $\mathbf{X} \otimes \vec{V}$ produces a 3-dimensional cubic grid $\vec{e}_\alpha \otimes \vec{e}_\beta \otimes \vec{e}_\gamma$ of all the triplets built by $\vec{e}_\alpha$ , $\vec{e}_\beta$ , $\vec{e}_\gamma$ and a similar cubic structure of all possible pairs $X^{\alpha\beta} T^\gamma$ .

▪ It should be emphasized that the dimension of the space on which the tensor extends is the rank $r$ of the tensor and has nothing to do with $n$, the dimension of geometric space. For instance, a rank 2 tensor is a matrix (2-dimensional) in a space of any dimension; what varies is the number $n$ of rows and columns.
Also the T-mosaic blockwise representation is invariant with the dimension of space: the number of connectors does not change since they are not individually represented, but as arrays.

▪ A physical example useful to make the notion of tensor more concrete can be the "stress tensor". It is a particular double tensor in the usual 3D space that plays a role in mechanics.

In a material body under stress we isolate a small tetrahedron *OABC* and look at the forces $\vec{s}, \vec{s}^{\,1}, \vec{s}^{\,2}, \vec{s}^{\,3}$ acting on each of its faces, respectively *ABC*, *OBC*, *OCA*, *OAB*, transmitted from the surrounding material (in the drawing the $x^1$ axis is to be seen enter into the sheet).

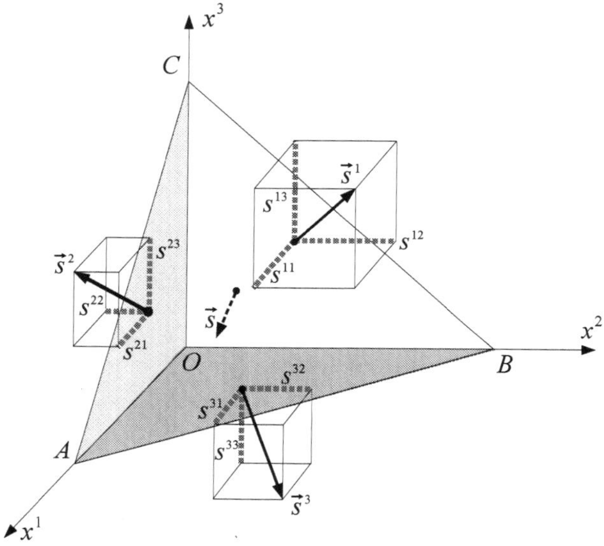


The figure shows the forces (stresses) acting on each face; $\vec{s}^{\,1}, \vec{s}^{\,2}, \vec{s}^{\,3}$ are decomposed into their respective components along coordinate axes. Static equilibrium implies $\vec{s}+\vec{s}^{\,1}+\vec{s}^{\,2}+\vec{s}^{\,3}=0$, so that $\vec{s}^{\,1}, \vec{s}^{\,2}, \vec{s}^{\,3}$ only, or the set of their components $s^{\alpha\beta}$, can describe the stress condition at the point.

The matrix $[s^{\alpha\beta}]$ represent an entity that enjoys tensor properties:

$$\mathbf{s} \equiv [s^{\alpha\beta}] = \begin{bmatrix} s^{11} & s^{12} & s^{13} \\ s^{21} & s^{22} & s^{23} \\ s^{31} & s^{32} & s^{33} \end{bmatrix} = \begin{bmatrix} \sigma^{11} & \tau^{12} & \tau^{13} \\ \tau^{21} & \sigma^{22} & \tau^{23} \\ \tau^{31} & \tau^{32} & \sigma^{33} \end{bmatrix},$$

the Cauchy stress tensor.

($\sigma$ are called normal stresses, $\tau$ shear stresses)

## 2.3 Sum of tensors and product by a number

As in the case of vectors and covectors, the set of all tensors of a certain rank $\binom{h}{k}$ defined at a point $P$ has the structure of a vector space, once defined operations of sum between tensors and multiplication of tensors by numbers.

The sum of tensors (of the same rank!) gives a tensor whose components are the sums of the components:

$$\mathbf{A}+\mathbf{B}=\mathbf{C} \quad \Leftrightarrow \quad C^{\alpha\beta\ldots}_{\mu\nu\ldots} = A^{\alpha\beta\ldots}_{\mu\nu\ldots} + B^{\alpha\beta\ldots}_{\mu\nu\ldots} \qquad 2.13$$

The product of a number $a$ by a tensor has the effect of multiplying by $a$ all the components:

$$a\mathbf{A} \overset{comp}{\rightarrow} a\, A^{\alpha\beta\ldots}_{\mu\nu\ldots} \qquad 2.14$$

## 2.4 Symmetry and skew-symmetry

A tensor is told ***symmetric*** with respect to a pair of indexes if their exchange leaves it unchanged; ***skew-symmetric*** if the exchange of the two indexes produces a change of sign (the two indexes must be *both upper or both lower indexes*).

For example: $T_{\alpha\beta\gamma\ldots}$ is

- symmetric with respect to indexes $\alpha, \gamma$ if $T_{\gamma\beta\alpha\ldots} = T_{\alpha\beta\gamma\ldots}$
- skew-symmetric with respect to indexes $\beta, \gamma$ if $T_{\alpha\gamma\beta\ldots} = -T_{\alpha\beta\gamma\ldots}$

Note that the symmetry has to do with the order of the arguments in the input list of the tensor. For example, taken a $\binom{0}{2}$ tensor:

$$\mathbf{T}(\vec{A},\vec{B}) = A^{\alpha} B^{\beta}\, \mathbf{T}(\vec{e}_{\alpha},\vec{e}_{\beta}) = A^{\alpha} B^{\beta} T_{\alpha\beta}$$

$$\mathbf{T}(\vec{B},\vec{A}) = B^{\beta} A^{\alpha}\, \mathbf{T}(\vec{e}_{\beta},\vec{e}_{\alpha}) = A^{\alpha} B^{\beta} T_{\beta\alpha}$$

$\Rightarrow$ the order of the arguments in the list is not relevant if and only if the tensor is symmetric, since:

$$T_{\alpha\beta} = T_{\beta\alpha} \quad \Leftrightarrow \quad \mathbf{T}(\vec{A},\vec{B}) = \mathbf{T}(\vec{B},\vec{A}) \qquad 2.15$$

▪ For tensors of rank $\binom{0}{2}$ or $\binom{2}{0}$, represented by a matrix, symmetry / skew-symmetry reflect into their matrices:

- symmetric tensor $\Rightarrow$ symmetric matrix: $[T_{\alpha\beta}] = [T_{\beta\alpha}]$
- skew-symmetric tensor $\Rightarrow$ skew-symmetric matrix $[T_{\alpha\beta}] = -[T_{\beta\alpha}]$

▪ Any tensor **T** ranked $\binom{0}{2}$ or $\binom{2}{0}$ can always be decomposed into a symmetric part **S** and skew-symmetric part **A** :

$$\mathbf{T} = \mathbf{S} + \mathbf{A}$$

that is $\qquad T_{\alpha\beta} = S_{\alpha\beta} + A_{\alpha\beta} \qquad$ where: $\qquad$ 2.16

$$S_{\alpha\beta} = \tfrac{1}{2}(T_{\alpha\beta} + T_{\beta\alpha}) \quad \text{and} \quad A_{\alpha\beta} = \tfrac{1}{2}(T_{\alpha\beta} - T_{\beta\alpha}) \qquad 2.17$$

Tensors of rank greater than 2 can have more complex symmetries with respect to exchanges of 3 or more indexes, or even groups of indexes.

▪ The symmetries are intrinsic properties of the tensor and do not depend on the choice of bases: if a tensor has a certain symmetry in a basis, it keeps the same in other bases (as it will become clear later).

## 2.5 Representing tensors in T-mosaic

▪ ***Tensor* T *"naked"***: it consists of a body that bears the indication of the components (in the shaded band); at edges the connectors:

- pins $\vec{e}_\alpha$ (in the top edge)
- holes $\tilde{e}^\mu$ (in the bottom edge)

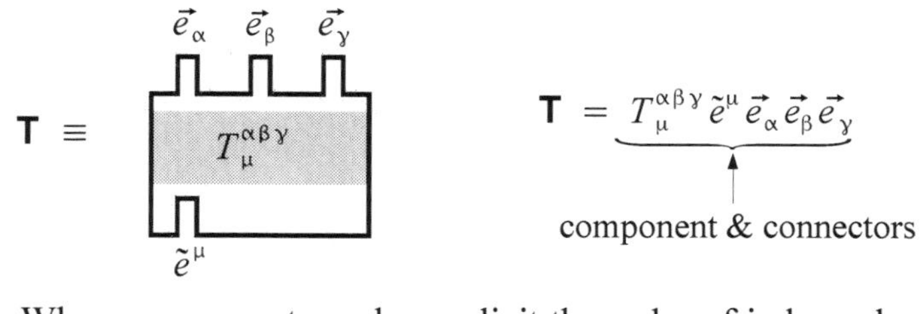


▫ When necessary to make explicit the order of indexes by means of reserved columns a representation by compartments will be used, as:

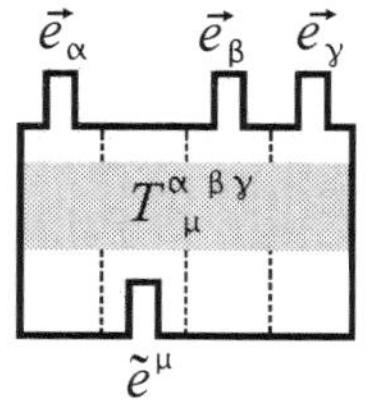


for $\mathbf{T} = T^{\alpha\ \beta\gamma}_{\ \mu}\,\vec{e}_\alpha\,\tilde{e}^\mu\,\vec{e}_\beta\,\vec{e}_\gamma$

The exposed connectors correspond to free indexes; they state the rank, given by $\binom{h}{k} = \binom{\text{number of pins } \vec{e}}{\text{number of holes } \tilde{e}}$ or even by $r = h + k$ .

▪ Blocks representing tensors can connect to each other with the usual rule: pins or holes of a tensor can connect with holes or pins (with equal generic index) of vectors / covectors, or other tensors. The meaning of each single connection is similar to that of the heterogeneous scalar product of vectors and covectors: connectors disappear and bodies merge *multiplying to each other.*

▪ When (at least) one of the factors is a tensor ($r > 1$) we properly refer to the product as *tensor inner product*.

▪ The shapes of the blocks can be deformed for graphic reasons without changing the order of the connectors. It is convenient to keep fixed the orientation of the connectors (pins "on", holes "down").

## 2.6 Tensors in T-mosaic model: definitions

Previously stated definitions for tensors and tensor components have simple graphical representation in T-mosaic as follows:

▪ ***Definition of tensor:*** plug all $r$ connectors of the tensor with vectors or covectors:

$$\mathbf{T}(\vec{A}, \tilde{P}, \tilde{Q}) = \mathbf{T}(A^\alpha \vec{e}_\alpha, P_\mu \tilde{e}^\mu, Q_\nu \tilde{e}^\nu) = A^\alpha P_\mu Q_\nu \ \mathbf{T}(\vec{e}_\alpha, \tilde{e}^\mu, \tilde{e}^\nu) =$$

$$= A^\alpha P_\mu Q_\nu T_\alpha^{\mu\nu} = X$$

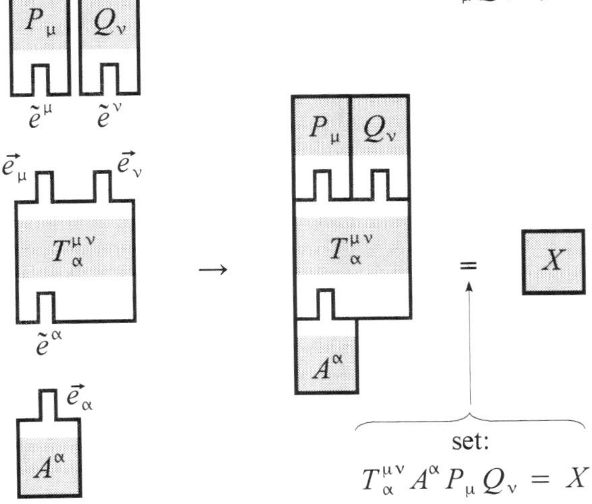

The result is a *single* number $X$ (all the indexes are dummy).

▪ ***Components of tensor* T** : produced by saturating with basis-vectors and basis-covectors all connectors of the "naked" tensor. For example:

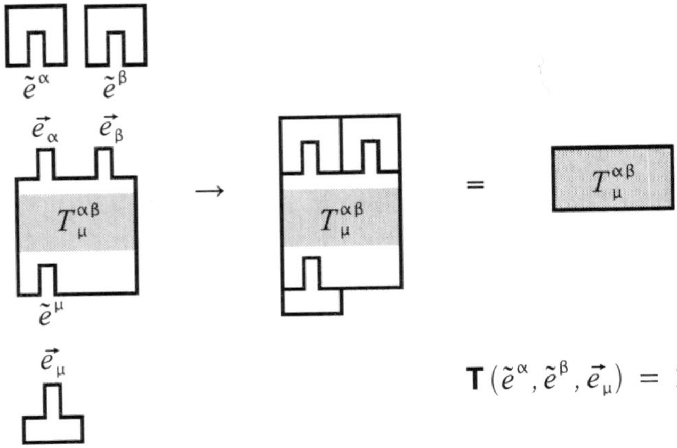


$$\mathbf{T}(\tilde{e}^{\alpha}, \tilde{e}^{\beta}, \vec{e}_{\mu}) = T^{\alpha\beta}_{\mu}$$

The result $T^{\alpha\beta}_{\mu}$ are $n^3$ numbers, the components (3 are the indexes).

Intermediate cases between the previous two can occur too:

▪ ***Input list with basis- and other vectors / covectors***:

$$\mathbf{T}(\vec{e}_{\alpha}, \tilde{P}, \tilde{e}^{\nu}) = \mathbf{T}(\vec{e}_{\alpha}, P_{\mu}\tilde{e}^{\mu}, \tilde{e}^{\nu}) = P_{\mu}\mathbf{T}(\vec{e}_{\alpha}, \tilde{e}^{\mu}, \tilde{e}^{\nu}) = P_{\mu}T^{\mu\nu}_{\alpha} = Y^{\nu}_{\alpha}$$

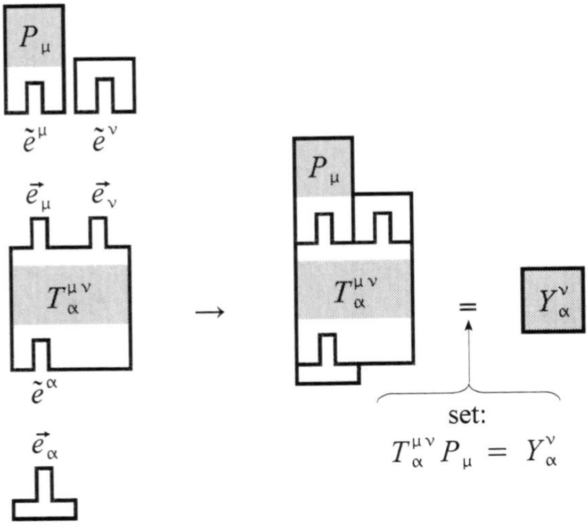

The result $Y^{\nu}_{\alpha}$ are $n^2$ numbers (survived indexes are 2): they are the components of a double tensor $\mathbf{Y} = Y^{\nu}_{\alpha}\vec{e}_{\nu}\tilde{e}^{\alpha}$; but it is improper to say that the result is a tensor!

*Remarks:*

- The order of connection, stated by the input list, is important. Note that a different result $\mathbf{T}(\vec{e}_{\alpha}, \tilde{e}^{\mu}, \tilde{P}) = Z^{\mu}_{\alpha} \neq \mathbf{T}(\vec{e}_{\alpha}, \tilde{P}, \tilde{e}^{\nu}) = Y^{\nu}_{\alpha}$ would be given by connecting $\tilde{P}$ with $\tilde{e}^{\nu}$ instead of $\tilde{e}^{\mu}$.

- In general a "coated" or "saturated" tensor, that is a tensor with all connectors plugged (by vector, covectors or other) so as to become a "smooth object" is a number or a multiplicity of numbers.

- It is worth to emphasize the different meaning of notations which are similar in appearance:

$\mathbf{T} = T^{\alpha\beta}_{\mu}\ \vec{e}_{\alpha}\vec{e}_{\beta}\tilde{e}^{\mu}$ is the tensor "naked"

$\mathbf{T}\underbrace{(\tilde{e}^{\alpha}, \tilde{e}^{\beta}, \vec{e}_{\mu})} = T^{\alpha\beta}_{\mu}$ is the result of applying the tensor to the list of basis-vectors (i.e. a component).

Input list:
"**T** applied to ..."

- The input list may also be *incomplete* (i.e. it may contain a number of arguments $< r$, rank of the tensor), so that some connectors remain unplugged. The application of a tensor to an incomplete list cannot result in a number or a set of number, but is a tensor lowered in rank.

## 2.7 Tensor inner product

The *tensor inner product* consists in the connection of two tensors **T** and **Y** realized by means of a pair of connectors (indexes) of different kind belonging one to **T**, the other to **Y**. The connection is *single* and involves only one $\vec{e}$ and only one $\tilde{e}$, but if the tensors are of high rank it can be realized in more than one way.

To denote the tensor inner product we will use the symbol $\bullet$ and we'll write $\mathbf{T}\bullet\mathbf{Y}$ to denote the product of the tensor **T** by **Y**, even if this writing can be ambiguous and requires further clarifications.

The two tensors that connect can be vectors, covectors or tensors of rank $r > 1$.

A particular case of tensor inner product is the already known heterogeneous scalar (or dot) product between a vector and a covector.

Let's define the total rank $R$ of the inner product as the sum of ranks of the two tensors involved: $R = r_1 + r_2$ ($R$ is the total number of connectors of the tensors involved in the product). For the heterogeneous scalar product is $R = 1 + 1 = 2$.

In a less trivial or strict sense the tensor inner product occurs between tensors of which at least one of rank $r > 1$, that is, $R > 2$. In any case, the tensor inner product lowers by 2 the rank $R$ of the result.

We will examine examples of tensor inner products of total rank $R$ gradually increasing, detecting their properties and peculiarities.

▪ ***Tensor inner product tensor • vector / covector***

For the moment we limit ourselves to the case $\mathbf{T} \cdot \vec{V}$ or $\mathbf{T} \cdot \tilde{P}$ for which $R = 2 + 1 = 3$.

Let's exemplify the case $\mathbf{T} \cdot \tilde{P}$. We observe that the tensor inner product of a rank $\binom{2}{0}$ tensor $\mathbf{T} = T^{\mu\nu} \vec{e}_\mu \vec{e}_\nu$ with a covector $P$ means applying the former to an incomplete list formed by the single element $\tilde{P}$, instead of a complete list like $\mathbf{T}(\tilde{P}, \tilde{Q})$.

We immediately see that this product $\mathbf{T} \cdot \tilde{P}$ can be made *in two ways*, depending on which one of the two connectors $\vec{e}$ of $\mathbf{T}$ is involved, and the results are in general different; we will distinguish them by writing $\mathbf{T}(\tilde{P},\ )$ or $\mathbf{T}(\ ,\tilde{P})$ where the space is for the missing argument. The respective schemes are:

$\mathbf{T}(\tilde{P},\ ) \equiv$

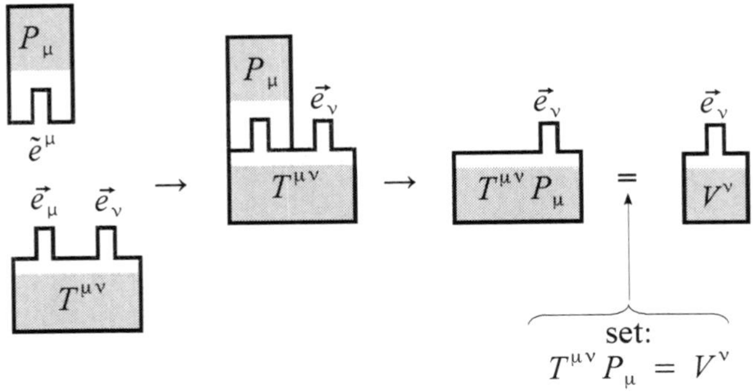


and:

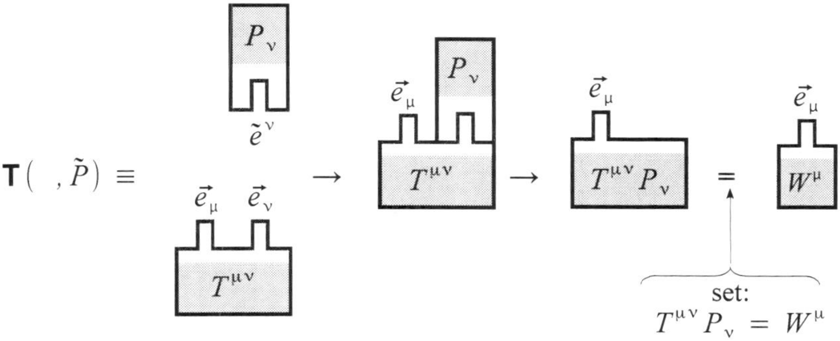


Only one connector of the tensor is plugged and then the rank $r$ of the tensor decrease by 1 (in the examples the result is a vector $\binom{1}{0}$ ).

Algebraic expressions corresponding to the two cases are:

$$\mathbf{T}\bullet\tilde{P} = T^{\mu\nu}\vec{e}_\mu\vec{e}_\nu(P_\alpha\,\tilde{e}^\alpha) = T^{\mu\nu}P_\alpha\vec{e}_\mu\vec{e}_\nu(\tilde{e}^\alpha) =$$

$$= \begin{cases} T^{\mu\nu}P_\alpha\underbrace{\vec{e}_\mu\vec{e}_\nu(\tilde{e}^\alpha)}_{\delta^\alpha_\mu} = T^{\mu\nu}P_\alpha\vec{e}_\nu\delta^\alpha_\mu = T^{\mu\nu}P_\mu\vec{e}_\nu = V^\nu\vec{e}_\nu \\ T^{\mu\nu}P_\alpha\vec{e}_\mu\underbrace{\vec{e}_\nu(\tilde{e}^\alpha)}_{\delta^\alpha_\nu} = T^{\mu\nu}P_\alpha\vec{e}_\mu\delta^\alpha_\nu = T^{\mu\nu}P_\nu\vec{e}_\mu = W^\mu\vec{e}_\mu \end{cases}$$

The notation $\mathbf{T}(\tilde{P})$ or $\mathbf{T}\bullet\tilde{P}$ may designate both cases, but it turns to be ambiguous because it does not specify the indexes involved in the inner product. Likewise ambiguous is the writing $\vec{e}_\mu\vec{e}_\nu(\tilde{e}^\alpha)$ in the equations written above.
In fact, in algebra as well as in T-mosaic block connection, we need to know which connectors plug (directly or by means of a Kronecker $\delta$) i.e. by which indexes the inner product is performed.
Only if $\mathbf{T}$ is symmetrical with respect to $\mu,\nu$ the result is the same and there is no ambiguity.

- In general, a writing like $\tilde{e}^\alpha\tilde{e}^\beta\vec{e}_\mu\vec{e}_\nu(\tilde{e}^\kappa)$ (which stands for $\tilde{e}^\alpha\otimes\tilde{e}^\beta\otimes\vec{e}_\mu\otimes\vec{e}_\nu(\tilde{e}^\kappa)$ with $\otimes$ implied) has the meaning of an inner product of the covector $\tilde{e}^\kappa$ by one element among those of the different kind aligned in the chain of outer products, for example:

$$\tilde{e}^\alpha\tilde{e}^\beta\underbrace{\vec{e}_\mu\vec{e}_\nu(\tilde{e}^\kappa)}_{\delta^\kappa_\mu} = \tilde{e}^\alpha\tilde{e}^\beta\vec{e}_\nu\delta^\kappa_\mu \quad \text{or} \quad \tilde{e}^\alpha\tilde{e}^\beta\vec{e}_\mu\underbrace{\vec{e}_\nu(\tilde{e}^\kappa)}_{\delta^\kappa_\nu} = \tilde{e}^\alpha\tilde{e}^\beta\vec{e}_\mu\delta^\kappa_\nu$$

The results are different depending on the element (the index) "hooked" to form a $\delta$, that must be known from the context (no matter the position of $\tilde{e}^{\kappa}$ or $\vec{e}_{\kappa}$ in the chain).

Likewise, when the inner product is made with a vector, as in $\tilde{e}^{\alpha}\,\tilde{e}^{\beta}\,\vec{e}_{\mu}\,\vec{e}_{\nu}(\vec{e}_{\kappa})$ we have two chances:

$$\underbrace{\tilde{e}^{\alpha}\,\tilde{e}^{\beta}\,\vec{e}_{\mu}\,\vec{e}_{\nu}(\vec{e}_{\kappa})}_{\delta^{\alpha}_{\kappa}} = \tilde{e}^{\beta}\,\vec{e}_{\mu}\,\vec{e}_{\nu}\,\delta^{\alpha}_{\kappa} \quad \text{or} \quad \tilde{e}^{\alpha}\,\underbrace{\tilde{e}^{\beta}\,\vec{e}_{\mu}\,\vec{e}_{\nu}(\vec{e}_{\kappa})}_{\delta^{\beta}_{\kappa}} = \tilde{e}^{\alpha}\,\vec{e}_{\mu}\,\vec{e}_{\nu}\,\delta^{\beta}_{\kappa}$$

depending on whether the dot product goes to hook $\tilde{e}^{\alpha}$ or $\tilde{e}^{\beta}$.

Formally, the inner product of a vector or covector $\vec{e}_{\kappa}$ or $\tilde{e}^{\kappa}$ addressed to a certain element (different in kind) inside a tensorial chain $\tilde{e}^{\alpha}\,\tilde{e}^{\beta}\,\vec{e}_{\mu}\,\vec{e}_{\nu}\cdots$ removes the "hooked" element from the chain and insert a $\delta$ with the vanished indexes. The chain welds itself with a ring less, without changing the order of the remaining ones.

Note that product $\tilde{P}\bullet\mathbf{T}$ on fixed indexes cannot give a result different from $\mathbf{T}\bullet\tilde{P}$. Hence $\tilde{P}\bullet\mathbf{T} = \mathbf{T}\bullet\tilde{P}$ and the tensor inner product is *commutattive* in this case.

▪ Similar is the case of tensor inner product tensor • basis-covector. Here as well, the product can be performed in two ways: $\mathbf{T}(\tilde{e}^{\mu})$ or $\mathbf{T}(\tilde{e}^{\alpha})$. We linger on the latter case only, using the same $\mathbf{T}$ of previous examples:

$$\mathbf{T}(\ \ ,\tilde{e}^{\alpha}) = \mathbf{T}\bullet\tilde{e}^{\alpha} = T^{\mu\nu}\,\vec{e}_{\mu}\,\underbrace{\vec{e}_{\nu}(\tilde{e}^{\alpha})}_{\delta^{\alpha}_{\nu}} = T^{\mu\nu}\,\vec{e}_{\mu}\,\delta^{\alpha}_{\nu} = T^{\mu\alpha}\,\vec{e}_{\mu}$$

which we draw as:

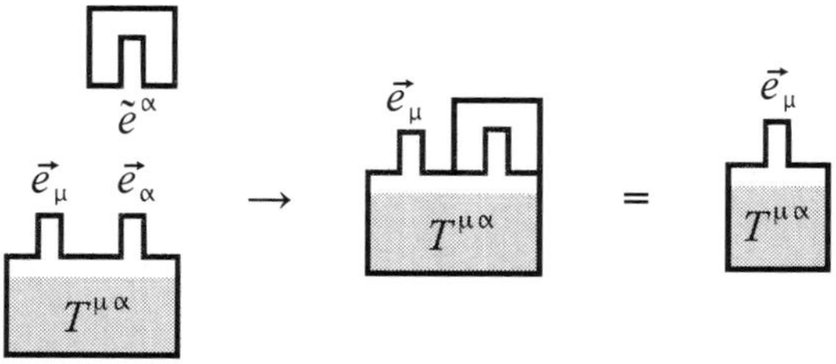

$T^{\mu\alpha}$, the component of $\mathbf{T}$, would be the result if the input list had been complete. Having omitted a basis-covector in the input list has the effect of leaving uncovered a connector $\vec{e}$, so that the result is a vector.

Similar considerations apply to tensor inner products tensor • vector where **T** is ranked $\binom{0}{2}$.

▪ If **T** is ranked $\binom{1}{1}$ the tensor inner products $\mathbf{T} \bullet \vec{V}$ and $\mathbf{T} \bullet \tilde{P}$ have an unique (not ambiguous) meaning because in both cases there is only one obliged way to perform them (it is clear in T-mosaic).
Here too the products commute: for example $\mathbf{T} \bullet \vec{V} = \vec{V} \bullet \mathbf{T}$.

▪ In conclusion: tensor inner products with rank $R = 3$ may or may not be twofold, but they are anyways commutative.

▪ ***Tensor inner product tensor • tensor*** $(R \geqslant 4)$
When the total rank $R$ of the tensor inner product is $R \geqslant 4$ the multiplicity of possible products increases and further complications arise, due to the non-commutativity of the products and for what concerns the order of the indexes that survive (when they come from both tensors, how order them?). Let's specify better.

▪ Given two tensors **A**, **B** we main by

• *multiplicity:* the number of inner products doable on all various indexes (connectors) *after fixed the order of the factors* (**A•B** or **B•A**).

• *commutativity:* the equality of "mirror-products", i.e. the eventuality that is **A • B** = **B • A** *after fixed the pair of indexes* (connectors) on which the products are performed.

The notion of commutativity applies to every single pair of mirror-products on pre-set indexes (= connectors).
The notion of multiplicity refers to a set of products, all the possible ones collected under the generic notation **A • B** (in a sense it is the measure of the ambiguity of the notation).

▪ The number of *different* products is less than multiplicity (= *possible* products) when **A**, **B** are somehow symmetrical: symmetries (i.e. the presence of equivalent interchangeable indexes) mean that some of the products give the same result.

▪ The multiplicity of **B•A** is clearly equal to that of **A•B** (there is 1 : 1 correspondence between each product and its mirrored).
The total number of *different* products doable with two tensors **A**, **B** (both as **A • B** and as **B • A**) is also reduced by the presence of commutative products: we have to consider that commutative pairs

should be counted as a *single* product instead of two.

In the following, we discuss a particular case of tensor inner product with $R = 4$, then we'll consider the general case of all possible products with $R = 4$.

▪ Example: a case $R=2+2=4$ is the following, with $\mathbf{A}\ \binom{0}{2}$ and $\mathbf{B}\ \binom{2}{0}$.

➊ *Multiplicity*

4 different tensor inner products $\mathbf{A} \bullet \mathbf{B}$ are possible, depending on which connectors plug (that is, which indexes are involved):

$$\mathbf{A} \bullet \mathbf{B} = \left(A_{\mu\nu} \tilde{e}^{\mu} \tilde{e}^{\nu}\right) \bullet \left(B^{\alpha\beta} \vec{e}_{\alpha} \vec{e}_{\beta}\right) =$$

$$= \begin{cases} \xrightarrow{\mu,\,\alpha} & = A_{\mu\nu} B^{\alpha\beta}\, \tilde{e}^{\nu} \vec{e}_{\beta}\, \delta^{\mu}_{\alpha} = A_{\alpha\nu} B^{\alpha\beta}\, \tilde{e}^{\nu} \vec{e}_{\beta} = C^{\beta}_{\nu}\, \tilde{e}^{\nu} \vec{e}_{\beta} \\ \xrightarrow{\nu,\,\alpha} & = A_{\mu\nu} B^{\alpha\beta}\, \tilde{e}^{\mu} \vec{e}_{\beta}\, \delta^{\nu}_{\alpha} = A^{\alpha\beta} B_{\mu\alpha} \tilde{e}^{\mu} \vec{e}_{\beta} = D^{\beta}_{\mu}\, \tilde{e}^{\mu} \vec{e}_{\beta} \\ \xrightarrow{\mu,\,\beta} & = A_{\mu\nu} B^{\alpha\beta}\, \tilde{e}^{\nu} \vec{e}_{\alpha}\, \delta^{\mu}_{\beta} = A^{\alpha\beta} B_{\beta\nu} \tilde{e}^{\nu} \vec{e}_{\alpha} = E^{\alpha}_{\nu} \tilde{e}^{\nu} \vec{e}_{\alpha} \\ \xrightarrow{\nu,\,\beta} & = A_{\mu\nu} B^{\alpha\beta}\, \tilde{e}^{\mu} \vec{e}_{\alpha}\, \delta^{\nu}_{\beta} = A^{\alpha\beta} B_{\mu\beta} \tilde{e}^{\mu} \vec{e}_{\alpha} = F^{\alpha}_{\mu} \tilde{e}^{\mu} \vec{e}_{\alpha} \end{cases}$$

which correspond respectively to the T-mosaic patterns: *

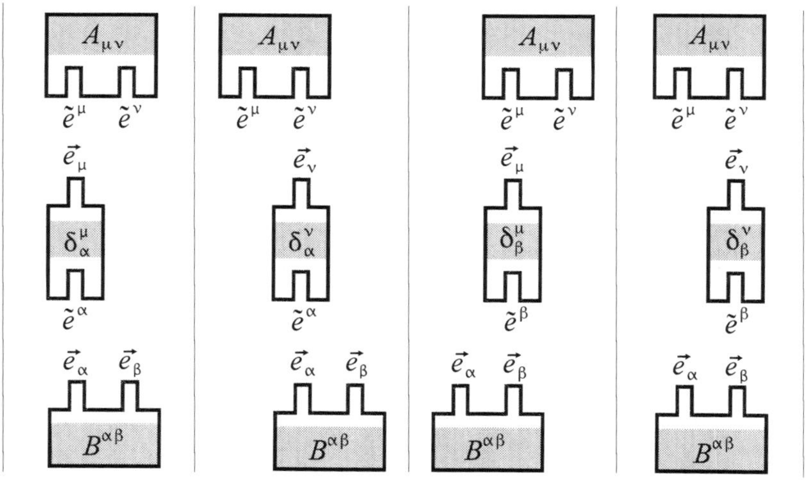

* We use here the artifice of δ as "plug adapter" to keep the correspondence of indexes between equations and blocks.

▪ The 4 results of **A•B** are in general different. But it may happen that:

**A** symmetrical $(\Rightarrow \tilde{e}^{\mu}, \tilde{e}^{\nu}$ interchangeable$) \Rightarrow \begin{cases} 1° \text{ product} = 2° \\ 3° \text{ product} = 4° \end{cases}$

**B** symmetrical $(\Rightarrow \vec{e}_{\alpha}, \vec{e}_{\beta}$ interchangeable$) \Rightarrow \begin{cases} 1° \text{ product} = 3° \\ 2° \text{ product} = 4° \end{cases}$

The 4 products coincide only if *both* **A** and **B** are symmetrical.

❷ *Non-commutativity and indexes' arrangement*

Mirrored product **B•A** stands for 4 more inner products, distinct from one another and different from the previous ones *; e.g. the first one is (product on $\alpha, \mu$):

$$\mathbf{B} \bullet \mathbf{A} = (B^{\alpha\beta} \vec{e}_{\alpha} \vec{e}_{\beta}) \bullet (A_{\mu\nu} \tilde{e}^{\mu} \tilde{e}^{\nu}) = C^{\beta}_{\nu} \vec{e}_{\beta} \tilde{e}^{\nu} \; .$$

Comparison with the first among the products above states that the tensor inner product on a given pair of indexes between tensors of rank $r = 2$ is *non-commutative*: **A•B** ≠ **B•A**.

▪ Note that in T-puzzle the graphic representation for **A•B** and **B•A** is the same. How do we account for the different results?
The distinction lies in a different reading of the result that takes account of the order of the factors and gives them the correct priority. In fact, the result must be constructed by the following rule: **

> list first surviving free connectors (indexes) of the first operand,
> then free connectors (indexes) survived from the second one.
> Reading of the connectors (indexes) is always from left to right.

Obviously indexes that make the connection don't appear in the result.

▪ If both **A** and **B** are symmetrical, then **A•B** = **B•A**, and the 4 + 4 inner products doable with **A** and **B** converge in one.

Ultimately, in the example we have found (leaving out for now the equally numerous mirror-products):

---

* Of course, the 4 possible pairs of indexes (connectors) are the same as the previous case, but their order is inverted. Note that the multiplicities of **A•B** are counted separately from those of **B•A**.

** This rule is fully consistent with the usual interpretation of the tensor inner product as "outer tensor product + contraction" which will be given later on.

• 4 products $\mathbf{A}\bullet\mathbf{B}$ with $\mathbf{A}\binom{0}{2}$ and $\mathbf{B}\binom{2}{0}$ (multiplicity 4), *non*-commutative.

▫ Going to consider the general case for $R = 4$ it is easy to see by means of T-puzzle that the set of all possible products includes, in addition to the 4 cases of the example, further cases:

• 7 products $\vec{V}\bullet\mathbf{T}$ with $\mathbf{T}\binom{0}{3},\binom{1}{2},\binom{2}{1}$ (multiplicity 3, 2, 1) all commutative;

• 7 products $\tilde{P}\bullet\mathbf{T}$ with $\mathbf{T}\binom{3}{0},\binom{2}{1},\binom{1}{2}$ (multiplicity 3, 2, 1) all commutative;

• 6 products $\mathbf{A}\bullet\mathbf{B}$ with $\mathbf{A}\binom{1}{1}$ and $\mathbf{B}\binom{2}{0},\binom{1}{1},\binom{0}{2}$ (multiplicity 2), *non*-commutative;

giving a total of 24 distinct inner products.
That means that 24 is the *multiplicity* for $R=4$.

The mirrored products made by exchanging the order of factors will also be 24, but partially coincident with the previous ones.

Including the mirrored products, the total number of different products will therefore be < 48 because of the pairs of commutating mirror-products which have to be counted as a single product instead of 2. Hence the correct count (in absence of symmetries in the tensors) will be, for the case R = 4:

$$4\times2+7\times1+7\times1+6\times2=34$$ different inner products.

▪ The number of doable products grows very rapidly with $R$; only the presence of symmetries in tensors can drastically reduce the number of different products.

▪ Let us note, finally, that the inner product of a vector or covector by a tensor of any rank always enjoys of the commutative property.

▪ ***Incomplete input lists and tensor inner product***

There is a partial overlap of the notions of inner product and the application of a tensor to an "input list". It is true that the complete input list is used to saturate all the connectors (indexes) of the tensor in order to give a number as a result; however, the list may be incomplete, and then one or more connectors (indexes) residues of the tensor are found in the result, which turns out to be a tensor of lowered rank. The extreme case of "incomplete input list" is just the tensor inner product tensor • vector or covector, where all the arguments of the list are missing, except one. Conversely the application of a tensor to a list (complete or not) can be seen as a succession of more than one single tensor inner products with vectors and covectors.

The higher the rank $r$ of the tensor, the larger the number of possible incomplete input lists, i.e. of situations intermediate between the complete list and the single tensor inner product with vector or covector. For example, for $r$=3 there are 6 possible cases of incomplete list. * In case of a $\binom{2}{1}$ tensor one among them is:

$\mathbf{T}(\ ,\tilde P,\tilde e^{\nu}) \equiv$

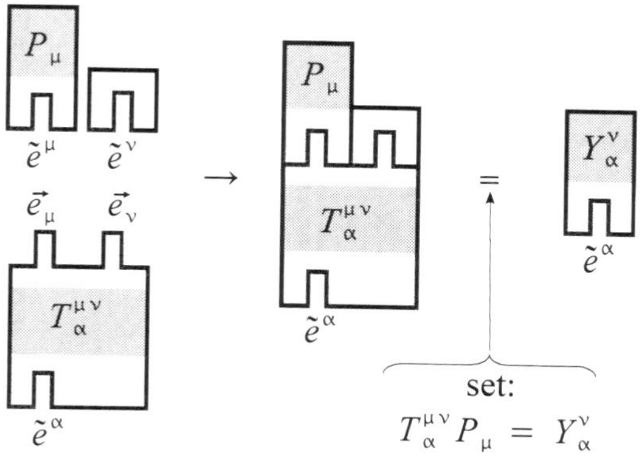


The result is $n$ covectors, one for each value of $\nu$.

This scheme can be interpreted as a tensor inner product between the tensor T and the covector $\tilde P$ , followed by a second tensor inner product between the result and the basis-covector $\tilde e^{\nu}$, treated as such in algebraic terms:

$$i)\quad \mathbf{T}(\tilde P) = T^{\mu\nu}_{\alpha}\,\tilde e^{\alpha}\underbrace{\vec e_{\mu}\,\vec e_{\nu}\,(P_{\kappa}\,\tilde e^{\kappa})}_{\delta^{\kappa}_{\mu}} = T^{\mu\nu}_{\alpha}\,P_{\kappa}\,\tilde e^{\alpha}\,\vec e_{\nu}\,\delta^{\kappa}_{\mu} = T^{\mu\nu}_{\alpha}\,P_{\mu}\,\tilde e^{\alpha}\,\vec e_{\nu} = Y^{\nu}_{\alpha}\,\tilde e^{\alpha}\,\vec e_{\nu}$$

$$ii)\quad Y^{\nu}_{\alpha}\,\tilde e^{\alpha}\underbrace{\vec e_{\nu}\,(\tilde e^{\gamma})}_{\delta^{\gamma}_{\nu}} = Y^{\nu}_{\alpha}\,\tilde e^{\alpha}\,\delta^{\gamma}_{\nu} = Y^{\gamma}_{\alpha}\,\tilde e^{\alpha}$$

* 3 lists lacking of one argument + 3 lists lacking of 2 arguments.

## 2.8 Outer product in T-mosaic

According to T-mosaic metaphor the tensor outer product $\otimes$ means "union by side gluing". It may involve vectors, covectors or tensors. Of the more general case is part the outer product between vectors and/or covectors; for example:

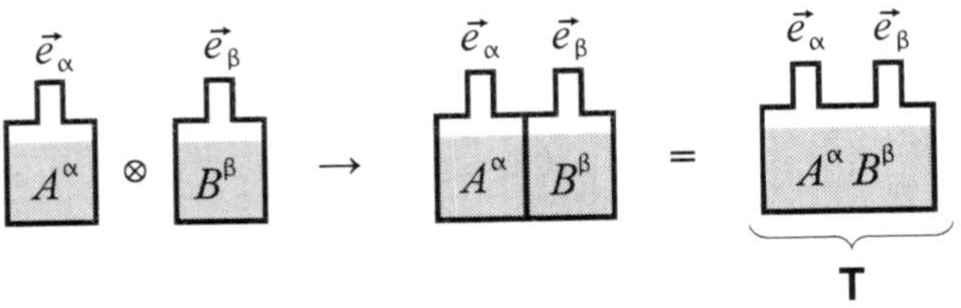

$$\vec{A} \otimes \vec{B} = A^{\alpha} B^{\beta} \vec{e}_{\alpha} \otimes \vec{e}_{\beta} = A^{\alpha} B^{\beta} \vec{e}_{\alpha} \vec{e}_{\beta} = T^{\alpha\beta} \vec{e}_{\alpha} \vec{e}_{\beta} = \mathbf{T}$$

or else:

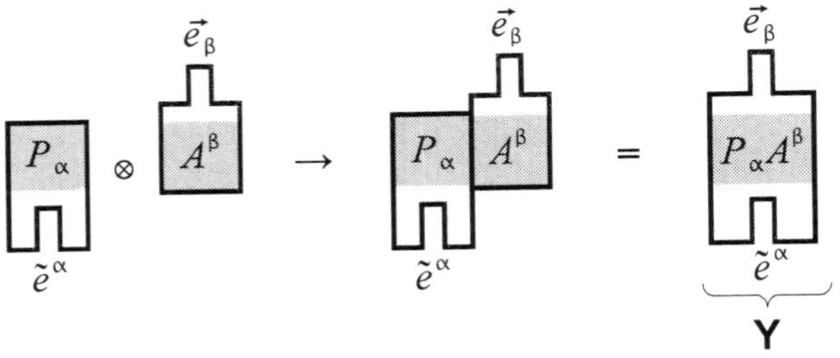

$$\tilde{P} \otimes \vec{A} = P_{\alpha} A^{\beta} \tilde{e}^{\alpha} \otimes \vec{e}_{\beta} = P_{\alpha} A^{\beta} \tilde{e}^{\alpha} \vec{e}_{\beta} = Y^{\beta}_{\alpha} \tilde{e}^{\alpha} \vec{e}_{\beta} = \mathbf{Y}$$

The outer product makes the bodies to blend together and components multiply (without the sum convention comes in operation).
It is clear from the T-mosaic representation that the symbol $\otimes$ has a meaning similar to that of the conjunction "and" in a list of items and can usually be omitted without ambiguity.

▪ The same logic applies to tensors of rank $r > 1$, and in such cases we speak of outer tensor product.
The outer tensor product operates on tensors of any rank, and merges them into one "by side gluing" The result is a composite tensor of rank equal to the sum of the ranks.
For example, in a case $\mathbf{A}\binom{1}{1} \otimes \mathbf{B}\binom{1}{2} = \mathbf{C}\binom{2}{3}$ :

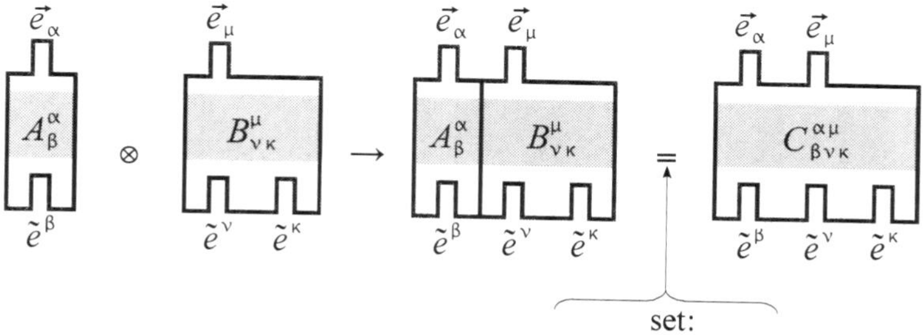

set:

$A^{\alpha}_{\beta}\, B^{\mu}_{\nu\kappa} = C^{\alpha\mu}_{\beta\nu\kappa}$

$$(A^{\alpha}_{\beta}\vec{e}_{\alpha}\otimes\tilde{e}^{\beta})\otimes(B^{\mu}_{\nu\kappa}\vec{e}_{\mu}\otimes\tilde{e}^{\nu}\otimes\tilde{e}^{\kappa}) = A^{\alpha}_{\beta}B^{\mu}_{\nu\kappa}\vec{e}_{\alpha}\otimes\tilde{e}^{\beta}\otimes\vec{e}_{\mu}\otimes\tilde{e}^{\nu}\otimes\tilde{e}^{\kappa} =$$
$$= C^{\alpha\mu}_{\beta\nu\kappa}\;\vec{e}_{\alpha}\otimes\tilde{e}^{\beta}\otimes\vec{e}_{\mu}\otimes\tilde{e}^{\nu}\otimes\tilde{e}^{\kappa}$$

usually written $C^{\alpha\mu}_{\beta\nu\kappa}\;\vec{e}_{\alpha}\,\tilde{e}^{\beta}\,\vec{e}_{\mu}\,\tilde{e}^{\nu}\,\tilde{e}^{\kappa}$ .

- Note the non-commutativity: $\mathbf{A}\otimes\mathbf{B}\neq\mathbf{B}\otimes\mathbf{A}$.

## 2.9 Contraction

Identifying two indexes of different kind (*one upper and one lower*), a tensor undergoes a *contraction* (the repeated index becomes dummy and appears no longer in the result). Contraction is an "unary" operation.

The tensor rank lowers from $\binom{h}{k}$ to $\binom{h-1}{k-1}$ .

In T-mosaic metaphor both a pin and a hole belonging to the same tensor disappear.

Example: contraction for $\alpha = \zeta$ :

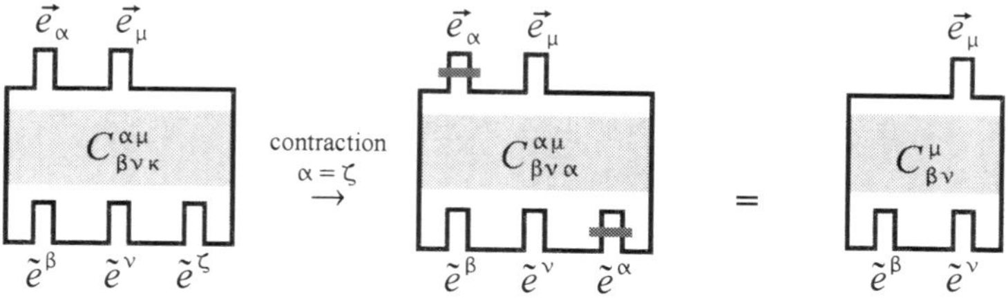


It is worth noting that the contraction is not a simple canceling or "simplification" of equal upper and lower indexes, but a sum which is triggered by repeated indexes. For example:

$$C^{\mu}_{\nu\mu} = C^{1}_{\nu 1} + C^{2}_{\nu 2} + \ldots + C^{n}_{\nu n} = C_{\nu}$$

▪ In a sense, the contraction consists of an inner product that operates inside the tensor itself, plugging two connectors of different kind.

▪ The contraction may be seen as the result of an inner product by a tensor carrying a connector (an index) which is already present, with opposite kind, in the operand. Typically this tensor is a Kronecker δ.

For example:

$C^{\alpha\mu}_{\beta\nu\kappa}\,\delta^{\kappa}_{\alpha} = C^{\alpha\mu}_{\beta\nu\alpha} = C^{\mu}_{\beta\nu}$ where contraction is on index $\alpha$.

The T-mosaic picture for the latter is the following:

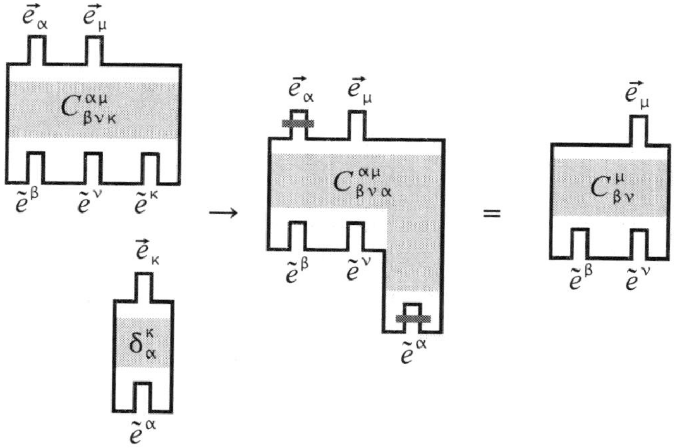

▪ For subsequent repeated contractions a tensor $\binom{h}{k}$ can be reduced until $h = 0$ or $k = 0$. A tensor of rank $\binom{h}{h}$ can be reduced to a scalar.

▪ The contraction of a $\binom{1}{1}$ tensor gives a scalar, the sum of the main diagonal elements of its matrix, and is called the ***trace***:

$$A^{\alpha}_{\alpha} = A^{1}_{1} + A^{2}_{2} + \ldots + A^{n}_{n} \tag{2.18}$$

The trace of **I** is $\delta^{\alpha}_{\alpha} = 1 + 1 + \ldots = n$ , dimension of the manifold. *

* **I** is the tensor "identity" whose components are Kronecker-δ : see later.

## 2.10 Inner product as outer product + contraction

The tensor inner product $\mathbf{A}\bullet\mathbf{B}$ can be interpreted as *tensor outer product* $\mathbf{A}\otimes\mathbf{B}$ *followed by a contraction of indexes*:

$$\mathbf{A}\bullet\mathbf{B} = contraction\ \ of\ (\mathbf{A}\otimes\mathbf{B}).$$

The multiplicity of possible contractions of indexes gives an account of the multiplicity of the tensor inner product represented by $\mathbf{A}\bullet\mathbf{B}$.

For example, in a case $\mathbf{A}\binom{0}{2}\bullet\mathbf{B}\binom{3}{0}=\mathbf{C}\binom{2}{1}$ , with total rank $R=5$:

$$\mathbf{A}\bullet\mathbf{B} = A_{\beta\nu}\cdot B^{\alpha\beta\gamma} = C_{\beta\nu}^{\cdot\,\cdot\,\alpha\beta\gamma} = C_{\nu}^{\cdot\,\alpha\gamma} \qquad \text{(product on } \beta\text{):}$$

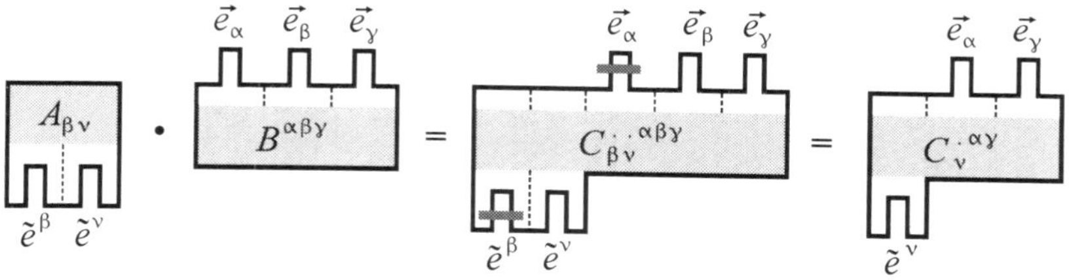


The 1st step (outer product) is univocal; the 2nd step (contraction) implies the choice of indexes (connectors) on which contraction must take place. Gradually choosing a connector after the other exhausts the choice of possibilities (multiplicity) of the inner product $\mathbf{A}\bullet\mathbf{B}$. The same considerations apply to the mirrored product $\mathbf{B}\bullet\mathbf{A}$.

▪ This 2-steps procedure is equivalent to the rules stated to write down the result of the tensor inner product. Its usefulness lies especially in complex cases such as:

$$A^{\mu\nu\cdot\zeta}_{\cdot\,\cdot\,\beta\,\cdot}\ B^{\cdot\,\beta\,\cdot}_{\alpha\,\cdot\,\gamma} = C^{\mu\nu\cdot\zeta\cdot\beta\cdot}_{\cdot\,\cdot\,\beta\,\cdot\,\alpha\,\cdot\,\gamma} = C^{\mu\nu\zeta\,\cdot\,\cdot}_{\cdot\,\cdot\,\cdot\,\alpha\gamma} \qquad 2.19$$

## 2.11 Multiple connection of tensors

In T-mosaic the tensor inner product is always performed by means of a simple connection between a pin $\vec{e}$ of a tensor and a hole $\tilde{e}$ of the other.
Multiple connections between two tensors, easy to fulfill in T-mosaic, can be algebraically described as an inner product followed by contractions in number of $m-1$, where $m$ is the number of plugged connectors) .

## 2.12 "Scalar product" or "identity" tensor

The operation of (heterogeneous) scalar product takes in input a vector and a covector to give a number $\Rightarrow$ it is a tensor of rank $\binom{1}{1}$. We denote it by $\mathbf{I}$:

$$\mathbf{I}(\tilde{P}, \vec{V}) \equiv \langle \tilde{P}, \vec{V} \rangle = P_\alpha V^\alpha \tag{2.20}$$

and expand it as: $\mathbf{I} = I^\alpha_\beta \; \vec{e}_\alpha \otimes \tilde{e}^\beta$, that is $\mathbf{I} = I^\alpha_\beta \, \vec{e}_\alpha \, \tilde{e}^\beta$.

How is it $\mathbf{I}$? Let us find its components giving to it the dual basis-vectors as input list: then

$$I^\alpha_\beta = \mathbf{I}(\tilde{e}^\alpha, \vec{e}_\beta) = \langle \tilde{e}^\alpha, \vec{e}_\beta \rangle = \delta^\alpha_\beta \tag{2.21}$$

by *eq.2.20*. Hence:

$$\mathbf{I} = \delta^\alpha_\beta \; \tilde{e}^\beta \vec{e}_\alpha \tag{2.22}$$

$\Rightarrow$ the Kronecker $\delta$ symbol has tensor character:

$$\boldsymbol{\delta} \equiv \mathbf{I} \equiv \left[\delta^\alpha_\beta\right] = \begin{bmatrix} \langle \tilde{e}^1, \vec{e}_1 \rangle & \langle \tilde{e}^1, \vec{e}_2 \rangle & \cdots & \langle \tilde{e}^1, \vec{e}_n \rangle \\ \langle \tilde{e}^2, \vec{e}_1 \rangle & \langle \tilde{e}^2, \vec{e}_2 \rangle & & \langle \tilde{e}^2, \vec{e}_n \rangle \\ \vdots & & & \vdots \\ \langle \tilde{e}^n, \vec{e}_1 \rangle & \langle \tilde{e}^n, \vec{e}_2 \rangle & \cdots & \langle \tilde{e}^n, \vec{e}_n \rangle \end{bmatrix} = \begin{bmatrix} 1 & 0 & \dots \\ 0 & 1 & \dots \\ \dots & \dots & 1 \end{bmatrix} \tag{2.23}$$

namely:

$$\boldsymbol{\delta} \equiv \mathbf{I} \overset{comp}{\rightarrow} \delta^\alpha_\beta \tag{2.24}$$

> The heterogeneous scalar product *is* the tensor $\mathbf{I}$, identity tensor
> its components are the Kronecker's $\delta^\alpha_\beta$
> it is represented by the unit matrix $I = diag\,(+1)$

How does $\mathbf{I}$ act? Its complete input list contains two arguments: a vector and a covector. Let's apply it to a partial list that contains only one of the two; there are 2 possibilities:

❶ $\mathbf{I}(\vec{V}) = \delta^\alpha_\beta \; \tilde{e}^\beta \vec{e}_\alpha (V^\gamma \vec{e}_\gamma) = \delta^\alpha_\beta \, V^\gamma \underbrace{\tilde{e}^\beta \vec{e}_\alpha (\vec{e}_\gamma)}_{\delta^\beta_\gamma} = \delta^\alpha_\beta \, \delta^\beta_\gamma V^\gamma \vec{e}_\alpha =$

$= \delta^\alpha_\gamma \vec{V}^\gamma \vec{e}_\alpha = V^\alpha \vec{e}_\alpha = \vec{V}$

❷ $\mathbf{I}(\tilde{P}) = \delta^\alpha_\beta \; \tilde{e}^\beta \vec{e}_\alpha (P_\gamma \tilde{e}^\gamma) = \delta^\alpha_\beta \, P_\gamma \tilde{e}^\beta \underbrace{\vec{e}_\alpha (\tilde{e}^\gamma)}_{\delta^\gamma_\alpha} = \delta^\alpha_\beta \, \delta^\gamma_\alpha P_\gamma \tilde{e}^\beta =$

$= \delta^\gamma_\beta P_\gamma \tilde{e}^\beta = P_\beta \tilde{e}^\beta = \tilde{P}$

(note: $\delta^\alpha_\beta \, \delta^\beta_\gamma = \delta^\alpha_\gamma$)

The T-mosaic blockwise representation of $\mathbf{I}(\vec{V})$ is:

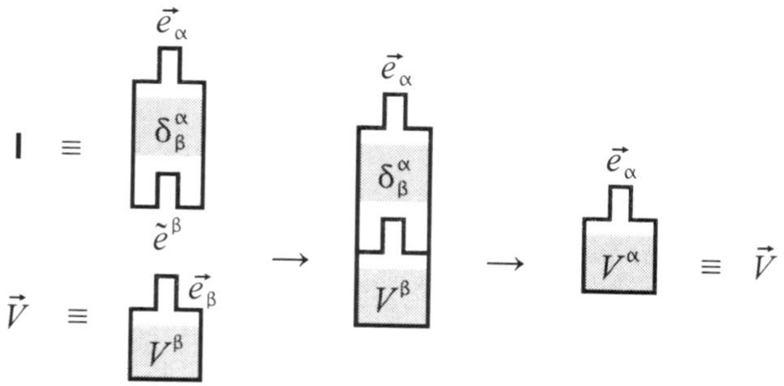


Actually, $\mathbf{I}$ transforms a vector into itself and a covector into itself:

$$\mathbf{I}(\vec{V}) = \vec{V} \quad ; \quad \mathbf{I}(\tilde{P}) = \tilde{P} \tag{2.25}$$

hence its name "identity tensor".

The same tensor is also called "fundamental mixed tensor".*

We also observe that, for $\forall \mathbf{T}$ :

$$\mathbf{T}(\mathbf{I}) = \mathbf{I}(\mathbf{T}) = \mathbf{T} \tag{2.26}$$

## 2.10 Inverse tensor

Given a tensor $\mathbf{T}$, if there exists a *single* tensor $\mathbf{Y}$ such that $\mathbf{T} \cdot \mathbf{Y} = \mathbf{I}$ , we say that $\mathbf{Y}$ is the inverse of $\mathbf{T}$, i.e. $\mathbf{Y} = \mathbf{T}^{-1}$ and

$$\mathbf{T} \cdot \mathbf{T}^{-1} = \mathbf{I} \tag{2.27}$$

Only tensors $\mathbf{T}$ of rank 2, and among them only $\binom{1}{1}$ type or *symmetric* $\binom{0}{2}$ or $\binom{2}{0}$ type tensors, can satisfy these conditions ** and have an inverse $\mathbf{T}^{-1}$.

---

* Hereafter the notation $\boldsymbol{\delta}$ will be abandoned in favor of $\mathbf{I}$.

** That *eq.2.27* is satisfied by tensors $\mathbf{T}$ and $\mathbf{T}^{-1}$ both ranked $r = 2$ can easily be shown in terms of T-mosaic blocks. If $\mathbf{T}$ e $\mathbf{T}^{-1}$ are tensors ranked $\binom{0}{2}$ or $\binom{2}{0}$ , *eq.2.27* stands for 4 different inner products and, for a given $\mathbf{T}$, can be satisfied for more than one $\mathbf{T}^{-1}$; only if we ask $\mathbf{T}$ to be symmetric ( $\Rightarrow \mathbf{T}^{-1}$ symmetric too) the tensor $\mathbf{T}^{-1}$ that satisfies it is unique and the uniqueness of the inverse is guaranteed. For this reason one restricts to symmetric tensors only.

For tensors of rank r $\neq$ 2 the inverse is not defined.

▪The inverse $\mathbf{T}^{-1}$ of a symmetric tensor $\mathbf{T}$ of rank $\binom{0}{2}$ or $\binom{2}{0}$ has the following properties:

- the indexes position interchanges upper $\leftrightarrow$ lower *
- it is represented by the inverse matrix
- it is symmetric

For example: given the $\binom{2}{0}$ symmetric tensor $\mathbf{T} \overset{comp}{\rightarrow} T^{\alpha\beta}$ , its inverse is the $\binom{0}{2}$ tensor $\mathbf{T}^{-1} \overset{comp}{\rightarrow} T_{\mu\nu}$ such that:

$$T^{\alpha\gamma}\, T_{\gamma\nu} = \delta^{\alpha}_{\nu} \qquad 2.28$$

Furthermore, denoting $T$ and $\overline{T}$ the matrices associated to $\mathbf{T}$ and $\mathbf{T}^{-1}$ , we have:

$$T \cdot \overline{T} = I \qquad 2.29$$

where $I$ is the unit matrix; namely, the matrices $T$ and $\overline{T}$ are inverse to each other.

▫ Indeed, for $\binom{2}{0}$ tensors the definition *eq.2.27* turns into *eq.2.28.* But the latter, written down in matrix terms, takes the form $\left[T^{\alpha\gamma}\right] \cdot \left[\overline{T}_{\gamma\nu}\right] = I$ , which is equivalent to *eq.2.29.*

Since the inverse of a symmetric matrix is symmetric as well, the symmetry of $\mathbf{T}^{-1}$ follows: if a tensor is symmetric, then its inverse is symmetric, too.

▪ The correspondence between the inverse matrix and the inverse tensor forwards to the tensor other properties of the inverse matrix:

- the commutativity $T \cdot \overline{T} = \overline{T} \cdot T = I$ , which applies to inverse matrices, is also true for inverse tensors: $\mathbf{T} \cdot \mathbf{T}^{-1} = \mathbf{T}^{-1} \cdot \mathbf{T} = \mathbf{I}$
- in order that $\mathbf{T}^{-1}$ exists, the inverse matrix must also exist and for that it is required to be $det\ T \neq 0$ . **

* This fact is often expressed by saying that the inverse of a double-contravariant tensor is a double-covariant tensor, and vice versa.

** A matrix can have an inverse only if its determinant is not zero.

Currently we say that tensors $T^{\alpha\beta}$ and $T_{\alpha\beta}$ are inverse to each other: usually we call the components of both tensors with the same symbol $T$ and distinguish them only by the position of the indexes. However, it is worth realizing that we are dealing with different tensors, and they cannot run both under the same symbol **T** (if we call **T** one of the two, we must use another name for the other: in this instance $\mathbf{T}^{-1}$).

▪ It should also be reminded that (only) if $T^{\alpha\beta}$ is diagonal (i.e. $T^{\alpha\beta} \neq 0$ only for $\alpha = \beta$ ) its inverse will be diagonal as well, with components $T_{\alpha\beta} = 1 / T^{\alpha\beta}$, and viceversa.

▪ The property of a tensor to have an inverse is intrinsic to the tensor itself and does not depend on the choice of bases: if a tensor has inverse in a basis, it has inverse in any other basis, too (as will become clear later on).

▪ An obvious property belongs to the mixed tensor $T^{\alpha}_{\beta}$ of rank $\binom{1}{1}$ "related" with both $T_{\alpha\beta}$ and $T^{\alpha\beta}$ defined by their inner product:

$$T^{\alpha}_{\beta} = T^{\alpha\gamma} T_{\gamma\beta}$$

Comparing this relation with *eq.2.28* (written as $T^{\alpha\gamma} T_{\gamma\beta} = \delta^{\alpha}_{\beta}$ with $\beta$ in place of $\nu$) we see that:

$$T^{\alpha}_{\beta} = \delta^{\alpha}_{\beta} \qquad 2.30$$

Indeed, the mixed fundamental tensor $\binom{1}{1}$ $\delta^{\alpha}_{\beta}$ , or tensor **I** , is the "common relative" of all couples of inverse tensors $T_{\alpha\beta}$ , $T^{\alpha\beta}$ ).*

▪ The mixed fundamental tensor $\mathbf{I} \xrightarrow{comp} \delta^{\alpha}_{\beta}$ has (with few others **) the property to be the inverse of itself.

▫ Indeed, an already noticed *** property of Kronecker's $\delta$:

$$\delta^{\alpha}_{\beta}\, \delta^{\beta}_{\gamma} = \delta^{\alpha}_{\gamma}$$

is the condition of inverse (similar to *eq.2.28*) for $\delta^{\alpha}_{\beta}$.
(T-mosaic icastically shows the meaning of this relation).

---

* That does not mean, of course, that it is the only existing mixed double tensor (think to $C^{\beta}_{\alpha} = A_{\alpha\gamma} B^{\gamma\beta}$ when **A** e **B** are not related to each other).

** Also auto-inverse are the tensors $\binom{1}{1}$ , whose matrices are mirror images of **I**.

***Already noticed while calculating $\mathbf{I}(\tilde{P})$.

## 2.14 Vector-covector "dual switch" tensor

A tensor of rank $\binom{0}{2}$ needs two vectors as input to give a scalar. If the input list is incomplete and consists in one vector only, the result is a covector:

$$\mathbf{G}(\vec{V}) = G_{\alpha\beta}\tilde{e}^{\alpha}\tilde{e}^{\beta}(V^{\gamma}\vec{e}_{\gamma}) = G_{\alpha\beta}V^{\gamma}\underbrace{\tilde{e}^{\alpha}\tilde{e}^{\beta}\bullet\vec{e}_{\gamma}}_{\delta^{\alpha}_{\gamma}} = G_{\alpha\beta}V^{\alpha}\tilde{e}^{\beta} = P_{\beta}\tilde{e}^{\beta} = \tilde{P}$$

after set $G_{\alpha\beta}V^{\alpha} = P_{\beta}$ .

The T-mosaic representation is:

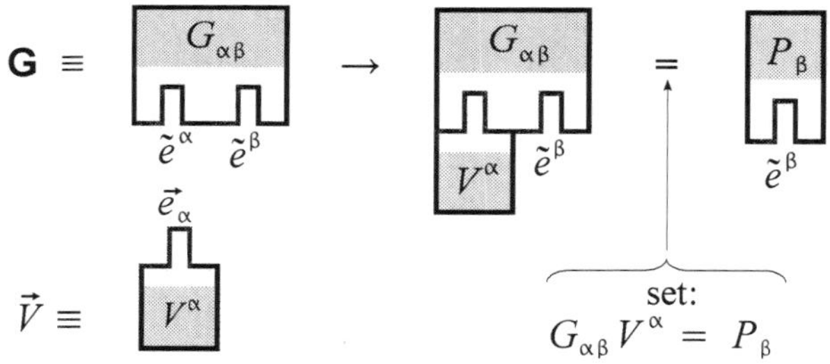


to be read: $\mathbf{G}(\vec{V}) = G_{\alpha\beta}V^{\alpha}\tilde{e}^{\beta} = P_{\beta}\tilde{e}^{\beta} = \tilde{P}$

By means of a $\binom{0}{2}$ tensor we can therefore transform a vector $\vec{V}$ into a covector $\tilde{P}$ that belongs to the dual space.

Let us pick out a $\binom{0}{2}$ tensor **G** as a "*dual switch*" to be used from now on to transform any vector $\vec{V}$ into its "dual" covector $\tilde{V}$ :

$$\mathbf{G}(\vec{V}) = \tilde{V} \qquad 2.29$$

**G** sets up a correspondence $\mathbf{G}: \vec{V} \rightarrow \tilde{V}$ between the two dual vector spaces (we use here the same name $V$ with different marks above in order to emphasize the relationship and the term "dual" as "related through **G** in the dual space").

The choice of **G** is arbitrary, nevertheless we must choose a tensor which has inverse $\mathbf{G}^{-1}$ in order to perform the "switching" in the opposite sense as well, from $\tilde{V}$ to $\vec{V}$ . In addition, if we want the inverse $\mathbf{G}^{-1}$ to be unique, we must pick out a **G** which is *symmetric*, as we know. Note that, at this point, using one or the other of the two $\tilde{e}$ connectors of **G** becomes indifferent.

Applying $\mathbf{G}^{-1}$ to the switch definition *eq.2.31*:

$$\mathbf{G}^{-1}(\mathbf{G}(\vec{V})) = \mathbf{G}^{-1}(\tilde{V})$$

but $\mathbf{G}^{-1}(\mathbf{G}) = \mathbf{I}$ and $\mathbf{I}(\vec{V}) = \vec{V}$ , then:

$$\vec{V} = \mathbf{G}^{-1}(\tilde{V}) \qquad 2.32$$

$\mathbf{G}^{-1}$ thus sets up an inverse correspondence $\mathbf{G}^{-1}: \tilde{V} \to \vec{V}$ .

In T-mosaic terms:

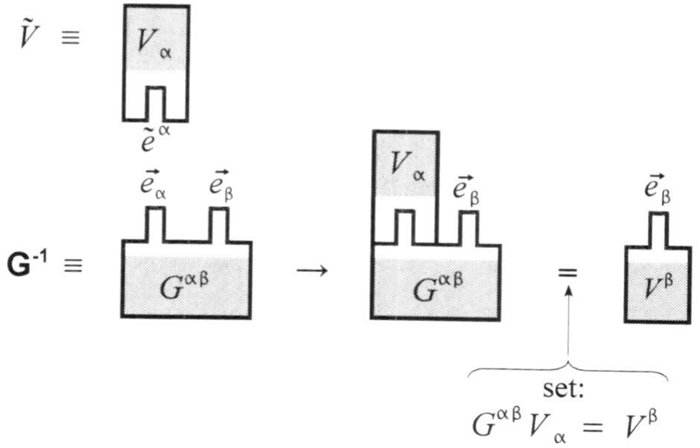


to be read $\mathbf{G}^{-1}(\tilde{V}) = G^{\alpha\beta} V_{\alpha}\, \vec{e}_{\beta} = V^{\beta} \vec{e}_{\beta} = \vec{V}$

▪ The vector ↔ covector switching can be expressed componentwise. In terms of components $\mathbf{G}(\vec{V}) = \tilde{V}$ is written as:

$$G_{\alpha\beta} V^{\alpha} = V_{\beta} \qquad 2.33$$

which is roughly interpreted: $G_{\alpha\beta}$ "***lowers the index***".

Conversely, $\mathbf{G}^{-1}(\tilde{V}) = \vec{V}$ can be written:

$$G^{\alpha\beta} V_{\alpha} = V^{\beta} \qquad 2.34$$

and interpreted as: $G^{\alpha\beta}$ "***raises the index***".

## 2.15 Vectors / covectors homogeneous scalar product

It presupposes the notion of switch tensor **G**.
We define the *homogeneous scalar product between two vectors* as the scalar product between one vector and the dual covector of the other:

$$\vec{A} \bullet \vec{B} \stackrel{\text{def}}{=} \langle \vec{A}, \tilde{B} \rangle \quad \text{or, likewise,} \quad \stackrel{\text{def}}{=} \langle \tilde{A}, \vec{B} \rangle \qquad 2.35$$

From the first equality, expanding on the bases and using the switch **G** we get:

$$\vec{A} \bullet \vec{B} = \langle \vec{A}, \tilde{B} \rangle = \langle A^{\alpha} \vec{e}_{\alpha}, B_{\gamma} \tilde{e}^{\gamma} \rangle = \langle A^{\alpha} \vec{e}_{\alpha}, \underbrace{G_{\beta\gamma} B^{\beta}}_{B_{\gamma}} \tilde{e}^{\gamma} \rangle =$$

$$= G_{\beta\gamma} A^{\alpha} B^{\beta} \langle \vec{e}_{\alpha}, \tilde{e}^{\gamma} \rangle = G_{\beta\gamma} A^{\alpha} B^{\beta} \delta_{\alpha}^{\gamma} = G_{\beta\gamma} A^{\gamma} B^{\beta} = \mathbf{G}(\vec{A}, \vec{B})$$

In short: $$\vec{A} \bullet \vec{B} = \mathbf{G}(\vec{A}, \vec{B}) \qquad 2.36$$

Hence, the dual switch tensor **G** is also the ***"scalar product between two vectors" tensor***.

The same result follows from the second equality *eq.2.35*.

The symmetry of **G** guarantees the commutative property of the scalar product:

$$\vec{A} \bullet \vec{B} = \vec{B} \bullet \vec{A}$$

or $$\mathbf{G}(\vec{A}, \vec{B}) = \mathbf{G}(\vec{B}, \vec{A}) \qquad 2.37$$

(note this is just the condition of symmetry for the tensor **G** *eq.2.15*).

▪ The *homogeneous scalar product between two covectors* is then defined by means of $\mathbf{G}^{-1}$:

$$\tilde{A} \bullet \tilde{B} = \mathbf{G}^{-1}(\tilde{A}, \tilde{B}) \qquad 2.38$$

$\mathbf{G}^{-1}$, the inverse dual switch, is thus the ***"scalar product between two covector" tensor***.

▪ In the T-mosaic metaphor the scalar (or inner) product between two vectors takes the form:

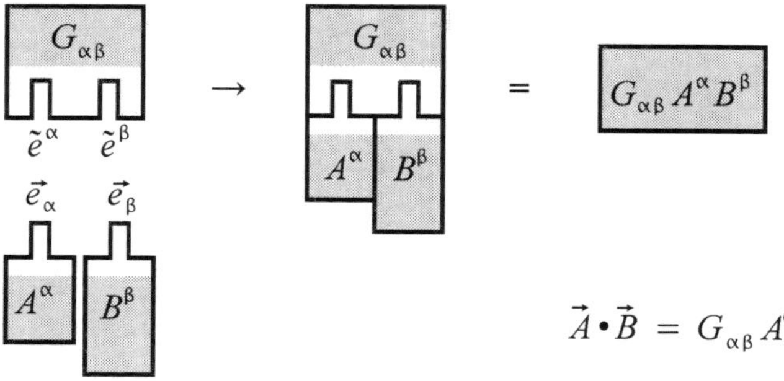


$$\vec{A} \bullet \vec{B} = G_{\alpha\beta} A^{\alpha} B^{\beta}$$

while the scalar (or inner) product between two covectors is:

$$\tilde{A} \bullet \tilde{B} = G^{\alpha\beta} A_\alpha B_\beta$$

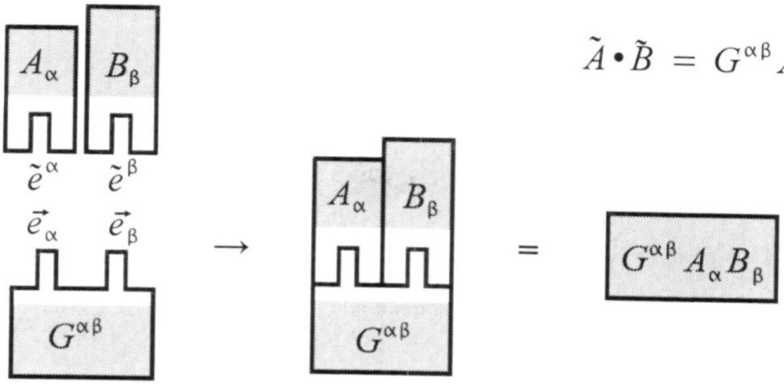


The result is a number in both cases.

▪ How is it **G**? Let's compute as usual its components using as arguments the basis-vectors instead of generic $\vec{A}$, $\vec{B}$ :

$$G_{\alpha\beta} = \mathbf{G}(\vec{e}_\alpha, \vec{e}_\beta) = \vec{e}_\alpha \bullet \vec{e}_\beta \qquad \Rightarrow$$

$$\Rightarrow \quad \mathbf{G} \equiv [G_{\alpha\beta}] = \begin{bmatrix} \vec{e}_1 \bullet \vec{e}_1 & \vec{e}_1 \bullet \vec{e}_2 & \cdots & \vec{e}_1 \bullet \vec{e}_n \\ \vec{e}_2 \bullet \vec{e}_1 & \vec{e}_2 \bullet \vec{e}_2 & & \vec{e}_2 \bullet \vec{e}_n \\ \vdots & & & \vdots \\ \vec{e}_n \bullet \vec{e}_1 & \vec{e}_n \bullet \vec{e}_2 & \cdots & \vec{e}_n \bullet \vec{e}_n \end{bmatrix} \qquad 2.39$$

It 's clear that the symmetry of **G** is related to the commutativity of the scalar product:

$$\vec{e}_\alpha \bullet \vec{e}_\beta = \vec{e}_\beta \bullet \vec{e}_\alpha \quad \Leftrightarrow \quad G_{\alpha\beta} = G_{\beta\alpha} \quad \Leftrightarrow \quad \mathbf{G} \text{ symmetric}$$

and that its associated matrix is also symmetric.

In a similar way for the inverse switch:

$$G^{\alpha\beta} = \mathbf{G}^{-1}(\tilde{e}^\alpha, \tilde{e}^\beta) = \tilde{e}^\alpha \bullet \tilde{e}^\beta \qquad \Rightarrow$$

$$\Rightarrow \quad \mathbf{G}^{-1} \equiv [G^{\alpha\beta}] = \begin{bmatrix} \tilde{e}^1 \bullet \tilde{e}^1 & \tilde{e}^1 \bullet \tilde{e}^2 & \cdots & \tilde{e}^1 \bullet \tilde{e}^n \\ \tilde{e}^2 \bullet \tilde{e}^1 & \tilde{e}^2 \bullet \tilde{e}^2 & & \tilde{e}^2 \bullet \tilde{e}^n \\ \vdots & & & \vdots \\ \tilde{e}^n \bullet \tilde{e}^1 & \tilde{e}^n \bullet \tilde{e}^2 & \cdots & \tilde{e}^n \bullet \tilde{e}^n \end{bmatrix} \qquad 2.40$$

*Notation*

Various equivalent writings for the homogeneous scalar product are:
- between vectors:
  $\vec{A}\bullet\vec{B} = \vec{B}\bullet\vec{A} = \langle \tilde{A}, \vec{B} \rangle = \langle \tilde{A}, \vec{B} \rangle = \mathbf{G}(\vec{A}, \vec{B}) = \mathbf{G}(\vec{B}, \vec{A})$
- between covectors:
  $\tilde{A}\bullet\tilde{B} = \tilde{B}\bullet\tilde{A} = \langle \tilde{A}, \vec{B} \rangle = \langle \vec{A}, \tilde{B} \rangle = \mathbf{G}^{-1}(\tilde{A}, \tilde{B}) = \mathbf{G}^{-1}(\tilde{B}, \tilde{A})$

The notation $\langle\ ,\ \rangle$ is reserved to heterogeneous scalar product vector-covector

## 2.16 G applied to basis-vectors

**G** transforms a basis-vector into a covector, but in general *not* into a basis-covector:

$$\mathbf{G}(\vec{e}_\alpha) = G_{\mu\nu}\tilde{e}^\mu \tilde{e}^\nu(\vec{e}_\alpha) = G_{\mu\nu}\tilde{e}^\mu \delta^\nu_\alpha = G_{\mu\alpha}\tilde{e}^\mu \qquad 2.41$$

and this is not $=\tilde{e}^\alpha$ because $G_{\mu\alpha}\tilde{e}^\mu$ is a sum over all $\tilde{e}^\mu$ and can't collapse to a single value $\tilde{e}^\alpha$ (except particular cases). Likewise:

$$\mathbf{G}^{-1}(\tilde{e}^\alpha) = G^{\mu\nu}\vec{e}_\mu \vec{e}_\nu(\tilde{e}^\alpha) = G^{\mu\nu}\vec{e}_\mu \delta^\alpha_\nu = G^{\mu\alpha}\vec{e}_\mu \neq \vec{e}_\alpha \qquad 2.42$$

▫ Notice that in *eq.2.41* we have made use of the duality condition (*eq.1.8*) that link the bases of vectors and covectors; it is a relation different from that stated by the “dual switching”.

## 2.17 G applied to a tensor

Can we assume that the converter **G** acts on the single index of the tensor (i.e. on the single connector $\vec{e}_\alpha$) as if the latter were isolated, in the same way it would act on a vector, that is changing it into a $\tilde{e}^\nu$ without altering other indices and their sequence? Not exactly. (The same question concerns its inverse $\mathbf{G}^{-1}$).

We see that the application of **G** to the tensor $\binom{3}{0}$ $\mathbf{X} = X^{\alpha\beta\gamma}\vec{e}_\alpha\vec{e}_\beta\vec{e}_\gamma$ :

$$G_{\alpha\nu}X^{\alpha\beta\gamma} = X_\nu^{\cdot\,\beta\gamma}$$

has, in this example, really the effect of lowering the index α involved in the product without altering the order of the remaining others. However, this is not the case for any index. What we need is to examine a series of cases more extensive than a single example, having in mind that the application of **G** to a tensor according to usual rules of the tensor inner product $\mathbf{G}(\mathbf{T}) = \mathbf{G}\bullet\mathbf{T}$ gives rise to a multiplicity of different products (the same applies to **G**$^{-1}$ ). To do so, let's think of the inner product as of the various possible contractions of the external product: in this interpretation it is up to the different contractions to produce multiplicity.

▫ Referring to the case, let's examine the other inner products we can do on the various indices, passing through the outer product $G_{\mu\nu}X^{\alpha\beta\gamma} = X_{\mu\nu}^{\cdot\cdot\alpha\beta\gamma}$ and then performing the contractions. The μ-α contraction leads to the known result $X_{\alpha\nu}^{\cdot\cdot\alpha\beta\gamma} = X_{\nu}^{\cdot\beta\gamma}$; in addition:

- contraction μ-β gives $X_{\beta\nu}^{\cdot\cdot\alpha\beta\gamma} = X_{\nu}^{\cdot\alpha\gamma}$
- contraction μ-γ gives $X_{\gamma\nu}^{\cdot\cdot\alpha\beta\gamma} = X_{\nu}^{\cdot\alpha\beta}$

(since **G** is symmetric, contractions of ν with α, β or γ give the same results as the contractions of μ).
Other results rise by the application of **G** in post-multiplication: from $X^{\alpha\beta\gamma}G_{\mu\nu} = X^{\alpha\beta\gamma}{}_{\cdot\cdot\cdot\mu\nu}$ we get $X^{\beta\gamma}{}_{\cdot\cdot\nu}$ , $X^{\alpha\gamma}{}_{\cdot\cdot\nu}$ , $X^{\alpha\beta}{}_{\cdot\cdot\nu}$ and other similar contractions for ν.
We observe that, among the results, there are lowerings of indices (e.g. $X_{\nu}^{\cdot\beta\gamma}$, lowering $^{\alpha}\searrow\nu$, and $X^{\alpha\beta}{}_{\cdot\cdot\nu}$ , lowering $^{\gamma}\searrow\nu$ ), together with other results that are *not* lowerings of indices (for example $X_{\nu}^{\cdot\alpha\gamma}$: β is lowered $^{\beta}\searrow\nu$ but also shifted). $X^{\alpha\cdot\gamma}{}_{\;\nu}$ does not appear among the results: it is therefore *not possible* to get by means of **G** the transformation:

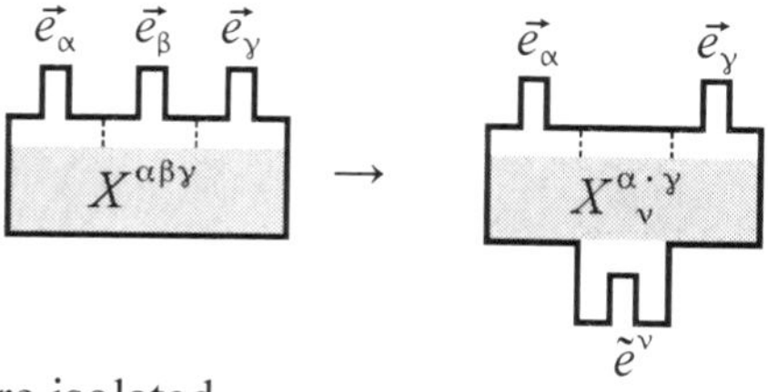


as if $\vec{e}_\beta$ were isolated.

It's easy to see that, given the usual rules of the inner tensor product, *only indices placed at the beginning or at the end of the string can be raised / lowered by* **G** (a similar conclusion applies to **G**$^{-1}$ ).
It is clear, however, that this limitation does not manifest itself until the indexes' string includes only two, that is, for tensors of rank $r=2$.

▪ For tensors of rank $r$=2 and, restricted to extreme indexes for tensors of higher rank, we can enunciate the rules:

**G** applied to a tensor ***lowers an index*** (the one by which it connects). Only the writing by components, e.g. $G_{\mu\nu} T^{\mu\alpha}_{\cdot\cdot\gamma} = T_{\nu\cdot\gamma}^{\cdot\alpha}$ , clarifies which indexes are involved.

**G**$^{-1}$ applied to a tensor ***raises an index*** (the one by which it connects). For example $T^{\alpha\beta}_{\cdot\cdot\gamma\mu} G^{\mu\nu} = T^{\alpha\beta\cdot\nu}_{\cdot\cdot\gamma}$ .

*Mnemo*

| | | |
|---|---|---|
| Examples: | $G_{\kappa\nu} T^{\kappa\alpha}_{\cdot\cdot\beta} = T_{\nu\cdot\beta}^{\cdot\alpha}$ : | hook $\kappa$, lower it, rename it $\nu$ |
| | $T^{\alpha\beta}_{\cdot\cdot\lambda\mu} G^{\mu\nu} = T^{\alpha\beta\cdot\nu}_{\cdot\cdot\lambda}$ | : hook $\mu$, raise it, rename it $\nu$ |

▪ A special case is when the tensor **G** applies to its inverse **G**$^{-1}$:

$$G_{\alpha\beta} G^{\beta\gamma} = G_{\alpha}^{\gamma}$$

which can be thought to as a raising as well as a lowering of an index. But since the two tensors **G** and **G**$^{-1}$ are inverse to each other:

$$G_{\alpha\beta} G^{\beta\gamma} = \delta_{\alpha}^{\gamma}$$

and from the last two it follows (not surprisingly, given the *eq.2.30*!) that:

$$G_{\alpha}^{\gamma} = \delta_{\alpha}^{\gamma} \qquad 2.43$$

## 2.18 Relations between I, G, δ

All equivalent to $\langle \tilde{A}, \vec{B} \rangle = \langle \vec{A}, \tilde{B} \rangle = \vec{A} \bullet \vec{B} = \tilde{A} \bullet \tilde{B}$ and to one another are the following expressions:

$$\mathbf{I}(\tilde{A}, \vec{B}) = \mathbf{I}(\vec{A}, \tilde{B}) = \mathbf{G}(\vec{A}, \vec{B}) = \mathbf{G}^{-1}(\tilde{A}, \tilde{B}) \qquad 2.44$$

(easy to ascertain using T-mosaic).

▪ The componentwise writing for $\mathbf{G}(\mathbf{G}^{-1}) = \mathbf{I}$ gives:

$$G_{\beta}^{\alpha} = I_{\beta}^{\alpha} \ ,$$

that, together with *eq.2.43*, leads to:

$$G_{\beta}^{\alpha} = I_{\beta}^{\alpha} = \delta_{\beta}^{\alpha} \qquad 2.45$$

Likewise:

$$G^{\alpha\beta} = I^{\alpha\beta} = \delta^{\alpha\beta} \ ; \ G_{\alpha\beta} = I_{\alpha\beta} = \delta_{\alpha\beta} \qquad 2.46$$

▫ That does *not* mean that **G** and **I** are the same tensor. We observe that the name **G** is reserved to the tensor whose components are $G_{\alpha\beta}$ ; the other two tensors, whose components are $G^{\alpha}_{\beta}$ and $G^{\alpha\beta}$ , are different tensors and cannot be labeled by the same name. In fact, the tensor with components $G^{\alpha\beta}$ is $\mathbf{G}^{-1}$, while that one whose components are $G^{\alpha}_{\beta}$ coincides with **I**.

It is thus matter of three distinct tensors: *

$$\begin{array}{lccccccc} \mathbf{I} & \stackrel{comp}{\rightarrow} & G^{\alpha}_{\beta} & = & I^{\alpha}_{\beta} & = & \delta^{\alpha}_{\beta} \\ \mathbf{G} & \stackrel{comp}{\rightarrow} & G_{\alpha\beta} & = & I_{\alpha\beta} & = & \delta_{\alpha\beta} \\ \mathbf{G}^{-1} & \stackrel{comp}{\rightarrow} & G^{\alpha\beta} & = & I^{\alpha\beta} & = & \delta^{\alpha\beta} \end{array}$$

even if the components' names are somewhat misleading.

Note that only $\delta^{\alpha}_{\beta}$ is the Kronecker delta, represented by the matrix *diag*(+1).

Besides, neither $I_{\alpha\beta}$ nor $I^{\alpha\beta}$ (and not even $\delta_{\alpha\beta}$ and $\delta^{\alpha\beta}$) deal with the identity tensor **I** (but rather with **G** which determine their form; their matrix may be *diag*(+1) or not).

The conclusion is that we can label by the same name tensors with indexes moved up / down when using a componentwise notation, but we must be careful, when switching to the tensor notation, to avoid identifying different tensors as one.

Just to avoid any confusion, in practice the notations $I_{\alpha\beta}$, $I^{\alpha\beta}$, $\delta_{\alpha\beta}$, $\delta^{\alpha\beta}$ are almost never used. Usually it is meant:

$$\begin{array}{lcc} \mathbf{I} & \stackrel{comp}{\rightarrow} & \delta^{\alpha}_{\beta} \\ \mathbf{G} & \stackrel{comp}{\rightarrow} & G_{\alpha\beta} \\ \mathbf{G}^{-1} & \stackrel{comp}{\rightarrow} & G^{\alpha\beta} \end{array}$$

* The rank is also different: $\binom{1}{1},\binom{0}{2},\binom{2}{0}$ respectively. Their matrix representation may be formally the same if $\mathbf{G} \equiv$ *diag* (+1), but on a *different "basis grid"!*

## 3 Change of basis

The vector $\vec{V}$ has expansion $V^{\alpha}\vec{e}_{\alpha}$ on the vector basis $\{\vec{e}_{\alpha}\}$.
Its components will change as the basis changes. In which way?
By the way, it's worth noting that the components change, but the vector $\vec{V}$ does not!
Let us denote by $V^{\beta'}$ the components on the new basis $\{\vec{e}_{\beta'}\}$ ; it is now $\vec{V} = V^{\beta'}\vec{e}_{\beta'}$ , hence:

$$\vec{V} = V^{\beta'}\vec{e}_{\beta'} = V^{\alpha}\vec{e}_{\alpha} \qquad 3.1$$

In other words, the same vector can be expanded on the new basis as well as it was expanded on the old one (from now on the upper $'$ will denote the new basis).

▪ Like all vectors, each basis-vector of the new basis $\vec{e}_{\beta'}$ can be expanded on the old basis-vectors $\vec{e}_{\alpha}$ :

$$\vec{e}_{\beta'} = \Lambda^{\alpha}_{\beta'}\,\vec{e}_{\alpha} \qquad (\beta' = 1,2,\ldots n) \qquad 3.2$$

$\Lambda^{\alpha}_{\beta'}$ are the coefficients of the expansion that describes the "recipe" of the new $\vec{e}_{\beta'}$ on the old basis $\{\vec{e}_{\alpha}\}$ (i.e. on the basis of the old "ingredients").
Taken as a whole, these coefficients express the new basis in terms of the old one; they can be arranged in a matrix $n \times n$ $\left[\Lambda^{\alpha}_{\beta'}\right]$

▪ *Conversely* we can express the old basis-vectors on the basis of new ones as:

$$\vec{e}_{\alpha} = \Lambda^{\beta'}_{\alpha}\,\vec{e}_{\beta'} \qquad (\alpha = 1,2,\ldots n) \qquad 3.3$$

The two matrices that appear in *eq.3.2* and *eq.3.3* are *inverse* to each other: exchanging the indexes the matrix is inverted because the sense of the transformation reverses; applying both them one after the other we return to the starting point.

▪ We now aim to find the relation between the new components of a given vector $\vec{V}$ and the old ones. Using *eq.3.3*:

$$\left.\begin{aligned} \vec{V} &= V^{\alpha}\vec{e}_{\alpha} = V^{\alpha}\Lambda^{\beta'}_{\alpha}\,\vec{e}_{\beta'} \\ \text{and since:}\quad \vec{V} &= V^{\beta'}\vec{e}_{\beta'} \end{aligned}\right\} \Rightarrow V^{\beta'} = \Lambda^{\beta'}_{\alpha}V^{\alpha} \qquad 3.4$$

▪ What about covectors? First let us deduce the transformation law for components. From $P_\alpha = \tilde{P}(\vec{e}_\alpha)$ *eq.1.4*, which also holds in the new basis, by means of the transformation of basis-vectors *eq.3.2* that we already know, we get:

$$P_{\beta'} = \tilde{P}(\vec{e}_{\beta'}) = \tilde{P}(\Lambda^\alpha_{\beta'}\vec{e}_\alpha) = \Lambda^\alpha_{\beta'}\tilde{P}(\vec{e}_\alpha) = \Lambda^\alpha_{\beta'}P_\alpha \qquad 3.5$$

Then, from the definition of component of $\tilde{P}$ and using *eq.3.5* above, we deduce the inverse transformation for basis-covectors (from new to old ones):

$$\left.\begin{array}{l}\tilde{P} = P_\alpha\,\tilde{e}^\alpha \\ \tilde{P} = P_{\beta'}\,\tilde{e}^{\beta'} = \Lambda^\alpha_{\beta'}\,P_\alpha\tilde{e}^{\beta'}\end{array}\right\} \Rightarrow \quad \tilde{e}^\alpha = \Lambda^\alpha_{\beta'}\,\tilde{e}^{\beta'}$$

and so the direct one from old to new:

$$\tilde{e}^{\beta'} = \Lambda^{\beta'}_\alpha\,\tilde{e}^\alpha \qquad 3.6$$

▪ To summarize, all direct transformations (from old to new basis) are ruled by two matrices:

$[\Lambda^{\beta'}_\alpha]$ expresses the transformation of the components of vectors (*eq.3.4*) and of basis-covectors (*eq.3.6*)

$[\Lambda^\alpha_{\beta'}]$ expresses the transformation of the components of covectors (*eq.3.5*) and of basis-vectors (*eq.3.2*)

The two matrices are one the transposed inverse * of the other.

---

* The transpose $M^T$ of a matrix $M$ is obtained by interchanging rows and columns; the transposed inverse $(M^{-1})^T$ coincides with the inverse transpose $(M^T)^{-1}$ . The notation we use cannot distinguish a matrix from its transpose (and the transposed inverse from the inverse, too) because the upper / lower indexes do not qualify rows and columns in a fixed way. Hence, the two matrices $\Lambda^{\beta'}_\alpha$ in *eq.3.3* and *eq.3.6* are not the same: they are the inverse in the first case and the transposed inverse in the second. It can be seen by expanding the two equations:

$$eq.3.3:\quad \vec{e}_\alpha = \Lambda^{\beta'}_\alpha\vec{e}_{\beta'} \Rightarrow \left\{\begin{array}{l}\vec{e}_1 = \Lambda^{1'}_1\vec{e}_{1'} + \Lambda^{2'}_1\vec{e}_{2'} + \dots \\ \vec{e}_2 = \Lambda^{1'}_2\vec{e}_{1'} + \Lambda^{2'}_2\vec{e}_{2'} + \dots \\ \dots\dots\dots\dots\end{array}\right\} \Rightarrow [\Lambda^{\beta'}_\alpha] = \begin{bmatrix}\Lambda^{1'}_1 & \Lambda^{2'}_1 & \cdots \\ \Lambda^{1'}_2 & \Lambda^{2'}_2 & \cdots \\ \cdots & & \end{bmatrix}$$

$$eq.3.6:\quad \tilde{e}^{\beta'} = \Lambda^{\beta'}_\alpha\,\tilde{e}^\alpha \Rightarrow \left\{\begin{array}{l}\tilde{e}^{1'} = \Lambda^{1'}_1\,\tilde{e}^1 + \Lambda^{1'}_2\,\tilde{e}^2 + \dots \\ \tilde{e}^{2'} = \Lambda^{2'}_1\,\tilde{e}^1 + \Lambda^{2'}_2\,\tilde{e}^2 + \dots \\ \dots\dots\dots\dots\end{array}\right\} \Rightarrow [\Lambda^{\beta'}_\alpha] = \begin{bmatrix}\Lambda^{1'}_1 & \Lambda^{1'}_2 & \cdots \\ \Lambda^{2'}_1 & \Lambda^{2'}_2 & \cdots \\ \cdots & & \end{bmatrix}$$

On the contrary, the inverse transformation (from old to new basis) is ruled by two matrices that are the inverse of the previous two.
So, just one matrix (with its inverse and transposed inverse) is enough to describe all the transformations of (components of) vectors and covectors, basis-vectors and basis-covectors under change of basis.

▪ Let us agree to denote $\Lambda$ (without further specification) *the matrix that transforms vector components and basis-covectors* from the old bases system with index $\alpha$ to the new system indexed $\beta'$:

$$\Lambda \stackrel{def}{=} \left[\Lambda_{\alpha}^{\beta'}\right] = \begin{bmatrix} \Lambda_1^{1'} & \Lambda_1^{2'} & \cdots \\ \Lambda_2^{1'} & \Lambda_2^{2'} & \cdots \\ \cdots & & \end{bmatrix} \qquad 3.7$$

Once set this position, the transformations of bases and components are summarized in the following table:

$$\begin{array}{ll} \vec{e}_{\alpha} \xrightarrow{\left(\Lambda^{-1}\right)^T} \vec{e}_{\beta'} & \tilde{e}^{\alpha} \xrightarrow{\Lambda} \tilde{e}^{\beta'} \\ V^{\alpha} \xrightarrow{\Lambda} V^{\beta'} & P_{\alpha} \xrightarrow{\left(\Lambda^{-1}\right)^T} P_{\beta'} \end{array} \qquad 3.8$$

Roughly speaking: if the bases vary in a way, the components vary in the opposite one in order to leave the vector unchanged (as well as the number that expresses the measure of a quantity increases by making smaller the unit we use, and viceversa).

▪ The transformations in the opposite sense ($\leftarrow$) require to invert all the matrices of the table (remind that the inverse of $\left(\Lambda^{-1}\right)^T$ is $\Lambda^T$ ).

▪ For a transformation to be reversible, it is required that its matrix $\Lambda$ is invertible, that means $det\,\Lambda \neq 0$ .

▪ The duality relation between bases of vectors and covectors holds in the new basis, too:

$$\langle \tilde{e}^{\nu'}, \vec{e}_{\beta'} \rangle = \langle \Lambda_{\mu}^{\nu'} \tilde{e}^{\mu}, \Lambda_{\beta'}^{\alpha} \vec{e}_{\alpha} \rangle = \Lambda_{\mu}^{\nu'} \Lambda_{\beta'}^{\alpha} \langle \tilde{e}^{\mu}, \vec{e}_{\alpha} \rangle =$$
$$= \Lambda_{\mu}^{\nu'} \Lambda_{\beta'}^{\alpha} \delta_{\alpha}^{\mu} = \Lambda_{\alpha}^{\nu'} \Lambda_{\beta'}^{\alpha} = \delta_{\beta'}^{\nu'}$$

(the two matrices are inverse to each other, $\left[\Lambda_{\mu}^{\nu'}\right] = \left[\Lambda_{\beta'}^{\alpha}\right]^{-1}$ and hence related by $\Lambda_{\alpha}^{\nu'} \Lambda_{\beta'}^{\alpha} = \delta_{\beta'}^{\nu'}$ ).

▪ The matrix $\Lambda$ that rules the transformations can be written in terms of both old and new basis. That can be seen by noting that:

$$\delta^{\mu}_{\alpha} = \langle \vec{e}_{\alpha}, \tilde{e}^{\mu} \rangle = \langle \vec{e}_{\alpha}, \Lambda^{\mu}_{\nu'} \tilde{e}^{\nu'} \rangle = \Lambda^{\mu}_{\nu'} \langle \vec{e}_{\alpha}, \tilde{e}^{\nu'} \rangle ,$$

possible only if $\langle \vec{e}_{\alpha}, \tilde{e}^{\nu'} \rangle = \Lambda^{\nu'}_{\alpha}$ because then $\delta^{\mu}_{\alpha} = \Lambda^{\mu}_{\nu'} \Lambda^{\nu'}_{\alpha}$ .
Hence, the element of the transformation matrix $\Lambda$ is:

$$\Lambda^{\nu'}_{\alpha} = \langle \vec{e}_{\alpha}, \tilde{e}^{\nu'} \rangle \qquad 3.9$$

built up crossing old and new bases.

▪ In practice, it's not convenient trying to remember the transformation matrices to use in each different case, nor reasoning in terms of matrix calculus: the balance of the indexes inherent in the sum convention is an automatic mechanism that leads to write the right formulas in every case. *

*Mnemo*

> The transformation of a given object is correctly written by simply taking care to balance the indexes. For example, the transformation of components of covector from new to old basis, provisionally written $P_{\mu} = \Lambda\ P_{\nu'}$ , can only be completed as $P_{\mu} = \Lambda^{\nu'}_{\mu} P_{\nu'}$ $\Rightarrow$ the matrix element to use here is $\Lambda^{\nu'}_{\mu}$

## 3.1 Basis change in T-mosaic

In T-mosaic metaphor, the change of basis requires *each connector to be converted.* To convert apply a "basis converter" extra-block to the connector; for convenience we represent it as a standard block with a pin and a hole (but beware: it is *not* a tensor!).

The basis-converter block is one and only one, in two variants distinguished by the different position of the indices with apex ′ :

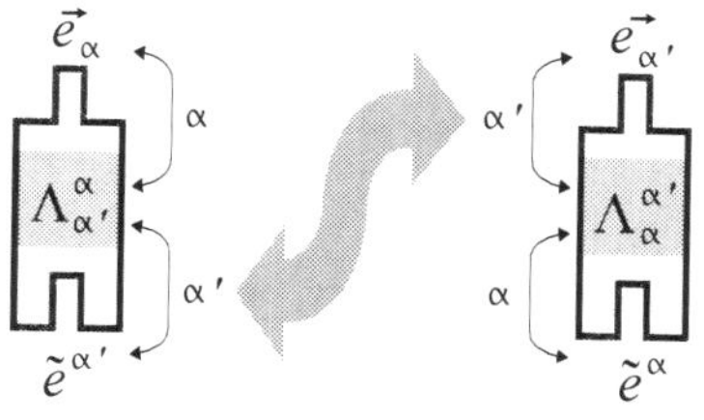


* We use indifferently e.g. $\Lambda^{\beta'}_{\alpha}$ or $\Lambda^{\alpha'}_{\alpha}$ entrusting to the apex the distinction.

The connection is made "wearing" the basis-converters blocks as "shoes" on the connectors of the tensor, docking them on the pin-side or the hole-side as needed, "apex on apex" or "non-apex on non-apex". The body of the converter blocks will be marked with the element of the transformation matrix $\Lambda^{\alpha'}_{\alpha}$ or $\Lambda^{\alpha}_{\alpha'}$ with apexes $'$ up or down turned in the same way as those of connectors (this implicitly leads to the correct choice of $\Lambda^{\alpha'}_{\alpha}$ or $\Lambda^{\alpha}_{\alpha'}$ ).

For example, to basis-transform the components of the vector $V^{\alpha}\vec{e}_{\alpha}$ we must put on the $\vec{e}_{\alpha}$ connector a converter block that plug up and replaces it with $\vec{e}_{\alpha'}$ :

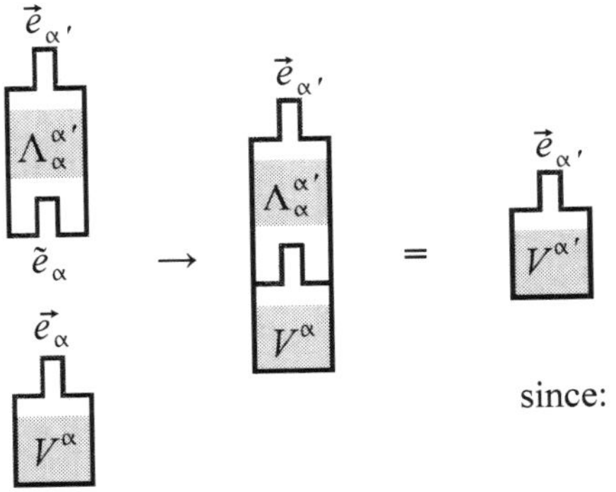


since: $\Lambda^{\alpha'}_{\alpha} V^{\alpha} = V^{\alpha'}$

Likewise, to basis-transform the components of the covector $P_{\alpha}\tilde{e}^{\alpha}$ , the appropriate converter block applies to the old connector $\tilde{e}^{\alpha}$ :

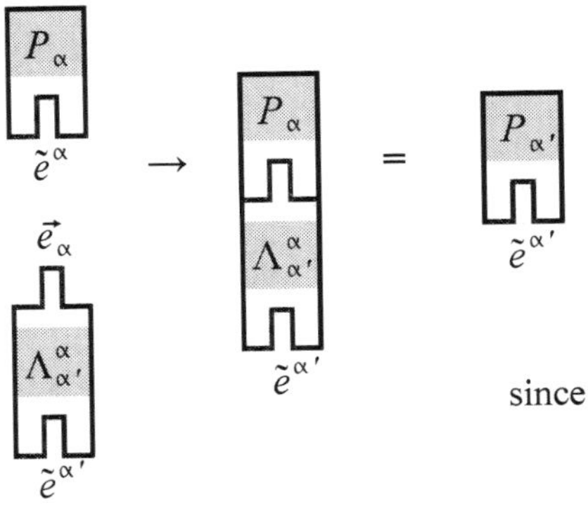


since: $\Lambda^{\alpha}_{\alpha'} P_{\alpha} = P_{\alpha'}$

▪ This rule doesn't apply to basis-vectors  (the converter block would contain in that case the inverse matrix, if one); however the block representation of these instances has no practical interest.

▪ Subject to a basis transformation, a tensor needs to convert ***all*** its connectors; an appropriate conversion block must be applied to each connector.

For example:

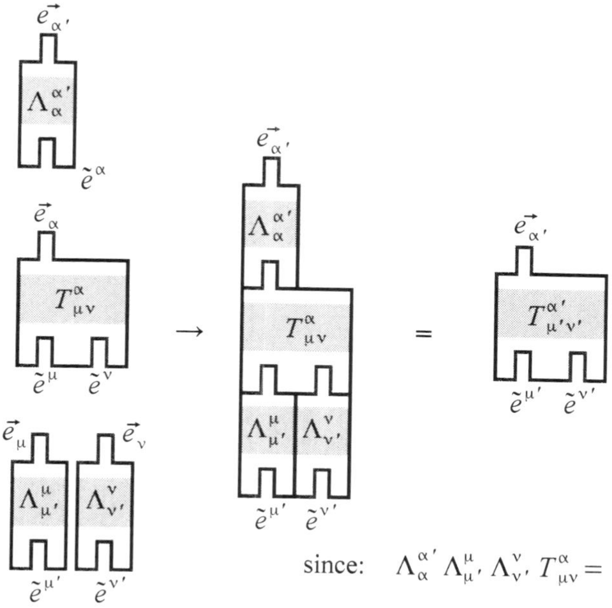

since: $\Lambda_{\alpha}^{\alpha'}\,\Lambda_{\mu'}^{\mu}\,\Lambda_{\nu'}^{\nu}\,T_{\mu\nu}^{\alpha} = T_{\mu'\nu'}^{\alpha'}$

The position just made:

$$T_{\mu'\nu'}^{\alpha'} = \Lambda_{\alpha}^{\alpha'}\,\Lambda_{\mu'}^{\mu}\,\Lambda_{\nu'}^{\nu}\,T_{\mu\nu}^{\alpha} \qquad 3.10$$

exemplifies the transformation rule for tensors:

> to old → new basis transform the components of a tensor we must apply as many $\Lambda_{\alpha}^{\alpha'}$ as upper indexes and as many $\Lambda_{\alpha'}^{\alpha}$ as lower indexes.

▪ A very special case is that of tensor $\mathbf{I} = \delta^{\alpha}_{\beta}\, \vec{e}_{\alpha}\, \tilde{e}^{\beta}$ whose components never change: $\delta^{\alpha}_{\beta} = diag(+1)$ is true ***in any basis***:

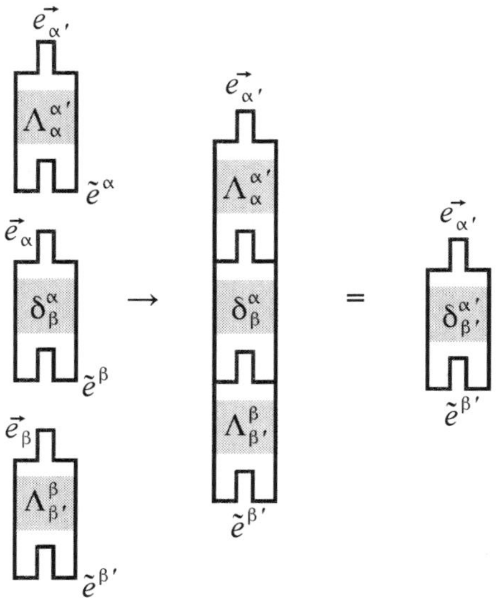


since: $\Lambda^{\alpha'}_{\alpha}\Lambda^{\beta}_{\beta'}\,\delta^{\alpha}_{\beta} = \Lambda^{\alpha'}_{\beta}\Lambda^{\beta}_{\beta'} = \delta^{\alpha'}_{\beta'}$

### 3.2 Invariance of the null tensor

A tensor is the null tensor when all its components are zero:

$$\mathbf{C} = 0 \quad \Leftrightarrow \quad \forall\, C^{\alpha\ldots}_{\mu\ldots} = 0 \quad * \qquad 3.11$$

If a tensor is null in a certain basis, it is also null in any other basis.

▫ In fact:

$$\mathbf{C} = 0 \;\Rightarrow\; C^{\alpha\ldots}_{\mu\ldots} = 0 \;\Rightarrow\; C^{\alpha'\ldots}_{\mu'\ldots} = \Lambda^{\alpha'}_{\alpha}\Lambda^{\mu}_{\mu'}\ldots C^{\alpha\ldots}_{\mu\ldots} = 0 \qquad 3.12$$

for all $\Lambda^{\alpha'}_{\alpha}, \Lambda^{\mu}_{\mu'}$ ... , that is for any transformation of bases.

From that the invariance of tensor equations follows.

---

* It makes no sense equaling a tensor to a number; writing $\mathbf{C} = 0$ is only a conventional notation for $\mathbf{C} = \mathbf{0}$ or $\forall\, C^{\alpha\ldots}_{\mu\ldots} = 0$.

## 3.3 Invariance of tensor equations

Vectors, covectors and tensors are the invariants of a given "landscape": changing the basis, only the way by which they are represented by their components varies. Then, also the relationships between them, or tensor equations, are invariant. For tensor equations we mean equations in which only tensors (vectors, covectors and scalars included) are involved, no matter whether written in tensor notation $(\mathbf{T}, \vec{V}, \text{ecc.})$ or componentwise.

Reduced to its essential terms, a tensor equation is an equality between two tensors like:

$$\mathbf{A} = \mathbf{B} \qquad \text{or} \qquad A^{\alpha\ldots}_{\mu\ldots} = B^{\alpha\ldots}_{\mu\ldots} \qquad 3.13$$

Now, it's enough to put $\mathbf{A} - \mathbf{B} = \mathbf{C}$ to reduce to the equivalent form *eq.3.12* $\mathbf{C} = 0$ or $C^{\alpha\ldots}_{\mu\ldots} = 0$ that, once valid in a basis, is valid in all bases. It follows that also *eq.3.13,* once valid in a basis, is valid in all bases. Hence the equations in tensor form are not affected by the particular basis chosen, but they hold in all bases without changes.
In practice

> an equation derived in a system of bases, once expressed in tensor form, applies as well in any system of bases.

Also some properties of tensors, if expressed by tensor relations, are valid regardless of the basis. It is the case, *inter alia*, of the properties of symmetry / skew-symmetry and invertibility. For example, *eq.2.15* that expresses the symmetry properties of a double tensor is a tensor equation, and that ensures that the symmetries of a tensor are the same in any basis.
The invertibility condition *eq.2.27*, *eq.2.28*, is a tensor equation too, and therefore a tensor which has inverse in a basis has inverse in any other basis.
As will be seen in the following paragraphs, the change of bases can be induced by a transformation of coordinates of the space in which tensors are set. Tensor equations, inasmuch invariant under change of basis regardless of the reason that this is due, will be *invariant under coordinate transformation* (that is, valid in any reference accessible via coordinate transformation).
In these features lies the strength and the greatest interest of the tensor formulation.

# 4 Tensors in manifolds

We have so far considered vectors and tensors defined at a single point $P$. This does not mean that $P$ should be an isolated point: vectors and tensors are usually given as *vector fields* or *tensor fields* defined in some domain or continuum of points.
Henceforth we will not restrict to $\mathbb{R}^n$ space, but we shall consider a wider class of spaces that retain some basic analytical properties of $\mathbb{R}^n$, such as the differentiability (of functions therein defined). Belongs to this larger class any $n$-dimensional space $\mathcal{M}$ whose points may be put in a one-to-one (= bijective) and continuous correspondence with the points of $\mathbb{R}^n$ (or its subsets). Continuity of correspondence means that points close in space $\mathcal{M}$ have as image points also close in $\mathbb{R}^n$, that is a requisite for the differentiability in $\mathcal{M}$.
Under these conditions we refer to $\mathcal{M}$ as a *differentiable manifold.*
(Here we'll also use, instead of the term "manifold", the less technical "space" with the same meaning).
Roughly speaking, a differentiable manifold of dimension $n$ is a space that can be continuously "mapped" in $\mathbb{R}^n$ (with the possible exception of some points).

▫ In more precise terms:

It is required that every infinitesimal neighborhood of the generic point $P \in \mathcal{M}$ has as image an infinitesimal neighborhood of the corresponding point $P'$ in $\mathbb{R}^n$ or, turning to finite terms, that for every open set $U \subset \mathcal{M}$ there is a continuous correspondence $\varphi\colon U \to \mathbb{R}^n$. A couple of elements $(U, \varphi)$ is called a "chart".

▪ One complication comes from the fact that, in general, there is not a correspondence that transports the whole space $\mathcal{M}$ into $\mathbb{R}^n$, i.e. there is not a single chart $(\mathcal{M}, \varphi)$. To map $\mathcal{M}$ completely we need a collection of charts $(U_i, \varphi_i)$, called "atlas".

▪ The charts of the atlas must have overlapping areas at the edges to ensure the transition between one chart to the other. Some points (or rather, some neighborhoods) will appear on more than one charts, each of which being the result of a different correspondence rule. For example, a point $Q$ will be mapped as point $\varphi(Q)$ on a chart and as $\psi(Q)$ on another one: it is required that the correspondence $\varphi(Q) \leftrightarrow \psi(Q)$ is itself continuous and differentiable.

An example of a differentiable manifold is a two-dimensional spherical surface like the Earth's surface, that can be covered by two-dimensional plane charts (even though no single chart can cover the whole globe and some points, different depending on the type of correspondence used, are left out; for example, the Mercator projection excludes the poles).

That there is not a single correspondence able to transport the whole manifold into $\mathbb{R}^n$ is not a heavy limitation when, as often occurs, only single points are left out. One correspondence can suffice in these cases to describe *almost* completely the space. In practical terms, we can say that there is a correspondence $\varphi$: $\mathcal{M} \rightarrow \mathbb{R}^n$ , meaning now with $\mathcal{M}$ the space deprived of some points.*

In any case is still valid the limitation that has brought us to consider so far vectors branching off from a single point: each vector or tensor is related to the point where it is defined and cannot be "operated" with vectors or tensors defined at different points, except they are infinitely close. This limitation comes out from the fact that we cannot presume to freely transport vectors (and tensors) in any space as it is usually done in the plane or in spaces $\mathbb{R}^n$.

As already mentioned, the set of tensors of rank $\binom{h}{k}$ defined at a point $P$ is itself a vector space of dimension $n^{h+k}$ . In particular, the already known vector space of dimension $n$ of the vectors defined at a point $P$ is called the *tangent space*; the similar for covectors is named the *cotangent space*.

Due to the regularity of the correspondence $\varphi$ , a small enough neighborhood of any point $P$ of the differentiable manifold will behave as a neighborhood of $\mathbb{R}^n$ : a differentiable manifold appears *locally* like a flat one, even if its overall structure is different.

In practice, the term differentiable manifold means a space to which we can apply a coordinate system: this is indeed the meaning of the

* In the following, when we talk about a coordinate system defined on a manifold, we will tacitly assume that some single points can be left outside. To be more precise, may be that "null measure sets" (as is a line in a 2D space) are excluded.

correspondence $\varphi$ .

## 4.1 Coordinate systems

Given a $n$-dimensional differentiable manifold $\mathcal{M}$ , defining a certain correspondence $\varphi$: $\mathcal{M} \rightarrow \mathbb{R}^n$ means "tagging" each point $P$ of $\mathcal{M}$ with $n$ numbers $x^\mu$ $(\mu = 1, 2, \dots n)$ that identify it uniquely. The $n$-tuple expresses the *coordinates* of $P$ (another way, punctual, to designate the correspondence $\varphi$ is to write $P \leftrightarrow \{x^\mu\}$ ).

Since the manifold is differentiable, the correspondence is continuous and transforms each neighborhood of $P$ into a neighborhood of $\{x^\mu\}$.

The numbers that express the coordinates may be lengths, angles or else. We note that a coordinate system does not presuppose any definition of distance between two points of the manifold.

## 4.2 Coordinate lines and surfaces

In the neighborhood of a point $P$ whose coordinates are $(x^1, x^2, \dots x^n)$ no other points can have the same coordinates, but there are points that have *some* coordinates equal to those of $P$ (provided that one at least is different).

The points that share with $P$ only a single coordinate form a (hyper) surface in $n-1$ dimensions. These ***coordinate surfaces*** passing through $P$ are $n$ in number, one for each coordinate which remains fixed.

The points that share with $P$ $n-1$ coordinates and differ only in one, let's call it $x^\nu$, align on a line along which only $x^\nu$ changes.

*n* ***coordinate lines*** of this kind branch out from $P$ .

A coordinate line originates from the intersection of $n-1$ coordinate (hyper)surfaces (the intersection leaves only one coordinate free). Thus, at each point $P$ of the manifold $n$ coordinate surfaces and $n$ coordinate lines intersect.

▫ The visualization comes easy in 3D:
At any point $P$ three coordinate surfaces intersect; on each of them the value of a coordinate remains constant. These surfaces, intersecting two by two, generate 3 coordinate lines; on each of them two coordinates are constant and only the third varies. The coordinate surfaces are labeled by the coordinate that remains constant, the coordinate lines by the coordinate that varies, as

shown in the figure.

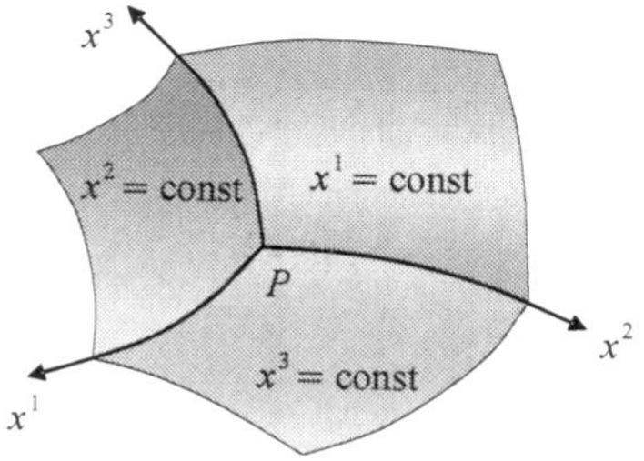


Roughly speaking: to define a coordinate system in a manifold means drawing an infinitely dense $n$-dimensional grid of coordinate lines (in general not straight nor intersecting at right angle) within the space.

## 4.3 Coordinate bases

An opportunity that arises once defined a coordinate system is to link the bases of vectors and covectors to the coordinates themselves.
The idea is to associate to the coordinates a basis of vectors consisting of $n$ basis-vectors, each of which tangent to one of the $n$ coordinate lines that intersect at the point.

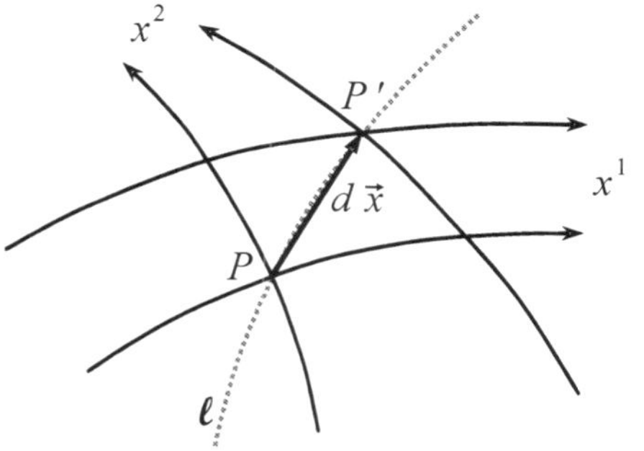


Given a generic point $P$ of coordinate $x^\mu$, infinite parametric lines * pass through it; let $\ell$ be one among them. Moving along $\ell$ by an infinitesimal the point moves from $P$ to $P'$ while the parameter

* A parametric line in a $n$-dimensional manifold is defined by a system of $n$ equations $x^\mu = x^\mu(s)$ with parameter $s$ ($\mu$ = 1, 2, ... $n$). To each value of $s$ corresponds a point of the line.

increases from $s$ to $s + ds$ and the coordinates of the point move from $x^1, x^2, \ldots x^n$ to $x^1 + dx^1, x^2 + dx^2, \ldots x^n + dx^n$.

The displacement from $P$ to $P'$ is then $d\vec{x} \overset{comp}{\longrightarrow} dx^\mu$, or:

$$d\vec{x} = dx^\mu \vec{e}_\mu . \qquad 4.1$$

This equation works as a definition of a vector basis $\{\vec{e}_\mu\}$ in $P$, in such a way that each basis-vector is tangent to a coordinate line.

Note that $d\vec{x}$ is a vector (inasmuch it is independent of the coordinate system which we refer to).

A basis of vectors defined in this way, tied to the coordinates, is called a ***coordinate vector basis***. Of course it is a basis among all the possible ones, but the easiest to use because only here $d\vec{x} \overset{comp}{\longrightarrow} dx^\mu$ *

It is worth to note that this basis is in general $P$-depending: $\vec{e}_\mu = \vec{e}_\mu(P)$.

▪ A further relation between $d\vec{x}$ and its components $dx^\mu$ is:

$$dx^\mu = \langle \tilde{e}^\mu, d\vec{x} \rangle \qquad 4.2$$

(it is nothing but the rule *eq.1.10*, according to which we get the components by applying the vector to the dual basis); its blockwise representation in T-mosaic metaphor is:

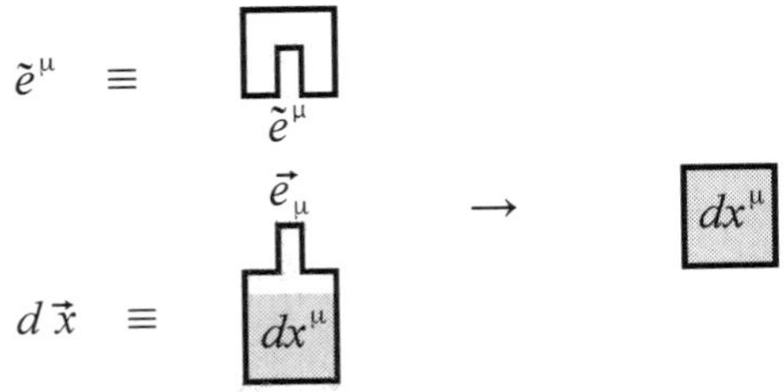


The basis $\{\tilde{e}^\mu\}$ is the ***coordinate covector basis,*** dual to the previously introduced coordinate vector basis.

▪ We aim now to link even this covector basis to the coordinates.

Let us consider a scalar function of the point $f$ defined on the manifold (at least along $\ell$) as a function of the coordinates. Its variation from the initial point $P$ along a path $d\vec{x}$ is given by its total differential:

$$df = \frac{\partial f}{\partial x^\mu} dx^\mu \qquad 4.3$$

---

* If the basis is not that defined by *eq.4.1*, we can still decompose $d\vec{x}$ along the coordinate lines $x^\mu$, but $dx^\mu$ is no longer a component along a basis-vector.

We note that the derivatives $\dfrac{\partial f}{\partial x^\mu}$ are the components of a covector because their product by components $dx^\mu$ of the vector $d\vec{x}$ gives the scalar *df* (in other words, *eq.4.3* can be interpreted as an heterogeneous scalar product).*

Let us denote by $\tilde{\nabla} f \overset{comp}{\rightarrow} \dfrac{\partial f}{\partial x^\mu}$ this covector, whose components *in a coordinate basis* are the partial derivatives of the function $f$, so that it can be identified with the gradient of $f$ itself.
In symbols, *in a coordinate basis*:

$$\tilde{grad} \;\equiv\; \tilde{\nabla} \;\overset{comp}{\rightarrow}\; \frac{\partial}{\partial x^\mu} \qquad 4.4$$

or, in short notation $\partial_\mu \equiv \dfrac{\partial}{\partial x^\mu}$ for partial derivatives:

$$\tilde{\nabla} \;\overset{comp}{\rightarrow}\; \partial_\mu \qquad 4.5$$

Note that the gradient is a covector, not a vector.

The total differential (*eq.4.3*) can now be written in vector form:

$$df = \tilde{\nabla} f\,(d\vec{x}) = \langle \tilde{\nabla} f,\, d\vec{x} \rangle$$

If as scalar function *f* we take one of the coordinates $x^\nu$ ** we get:

$$dx^\nu = \langle \tilde{\nabla} x^\nu, d\vec{x} \rangle$$

By comparison with the already known $dx^\nu = \langle \tilde{e}^\nu, d\vec{x} \rangle$ (*eq.4.2*) $\Rightarrow$

$$\Rightarrow \qquad \tilde{e}^\nu = \tilde{\nabla} x^\nu \quad \textit{in a coordinate basis} \qquad 4.6$$

This means that in coordinate bases the basis-covector $\tilde{e}^\nu$ coincides with the gradient of the coordinate $x^\nu$ and is therefore oriented in direction of the faster variation of the coordinate itself, which is the direction normal to the coordinate surface $x^\nu = const$ .

> In summary, provided we use *coordinate bases*:
> • the basis-vectors $\vec{e}_\mu$ are tangent to coordinate lines
> • the basis-covectors $\tilde{e}^\nu$ are normal to coordinate surfaces

---

* Note that, as we ask $dx^\mu$ to be a component of $d\vec{x}$, we assume to work *in a coordinate basis*.

** It means taking as line $\ell$ the coordinate line $x^\nu$.

as shown in the following figures that illustrate the 3D case:

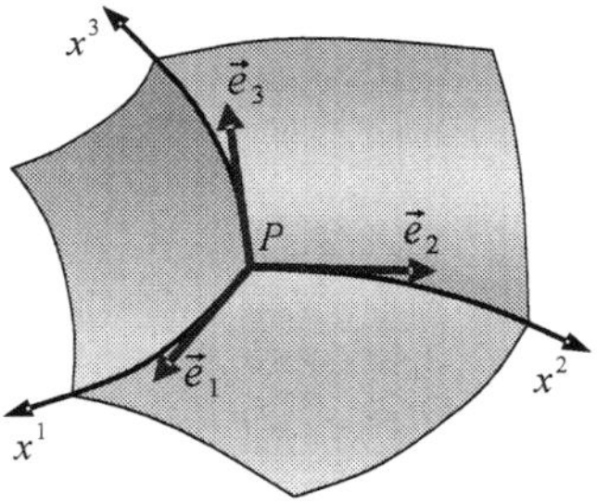


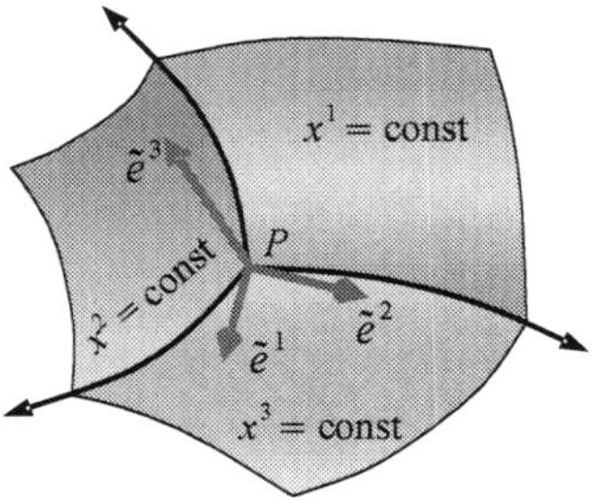


When the coordinate system is orthogonal ( $\Rightarrow$ the coordinate lines, in general not straight, intersect at each point at right angle) basis-vectors and basis-covectors have the same direction.* If the coordinates are not orthogonal the two sets of vectors and covectors are differently oriented.

Covectors, as well as vectors, can be represented by arrows, although we must keep in mind that they are entities of a different kind (for instance, vectors and covectors are not directly summable).**

## 4.4 Coordinate bases and non-coordinate bases

In 3D rectangular coordinates it is usual to write the infinitesimal displacement as:

$$d\vec{x} = \vec{i}\,dx + \vec{j}\,dy + \vec{k}\,dz$$

or, labeling by $\hat{e}$ the *unit* vectors:

$$d\vec{x} = \hat{e}_1 dx^1 + \hat{e}_2 dx^2 + \hat{e}_3 dx^3 = \hat{e}_\mu dx^\mu \qquad 4.7$$

which is just the condition of coordinate basis $d\vec{x} = \vec{e}_\mu dx^\mu$ *(eq.4.1)* provided $\vec{e}_\mu = \hat{e}_\mu$ .

This means that unit vectors $\hat{e}_\mu$, or $\vec{i}, \vec{j}, \vec{k}$, are a coordinate basis. Vectors $\vec{i}, \vec{j}, \vec{k}$, are in fact everywhere oriented as coordinate lines, and they are the same in the whole space $\mathbb{R}^3$ .

That is not always the case. For example, in plane polar coordinates

---

* In general, however, they will not match in length or magnitude.

** Some authors shy away from this representation, that is entirely lawful with the warning of the diversity of kind.

$(\rho, \theta)$ $d\vec{x}$ is given by:

$$d\vec{x} = \hat{e}_\rho d\rho + \hat{e}_\theta \rho d\theta \qquad 4.8$$

that is compatible with the coordinate basis condition $d\vec{x} = \vec{e}_\mu dx^\mu$ provided we take as a basis:

$$\vec{e}_\rho = \hat{e}_\rho \ , \quad \vec{e}_\theta = \rho\, \hat{e}_\theta \qquad 4.9$$

Here the basis-vector $\vec{e}_\theta$ "includes" $\rho$, it's no longer a unit vector and also varies from point to point. Conversely, the basis of unit vectors $\{\hat{e}_\rho, \hat{e}_\theta\}$ which is currently used in Vector Analysis for polar coordinates is *not* a coordinate basis.

▫ An example of a non-coordinate basis in 3D rectangular coordinates is obtained by applying a 45° rotation to unit vectors $\vec{i}, \vec{j}$ in the horizontal plane, leaving $\vec{k}$ unchanged.
Writing $\vec{i}, \vec{j}$ in terms of the new rotated vectors $\vec{i}\,', \vec{j}\,'$ we get for $d\vec{x}$:

$$d\vec{x} = \frac{\sqrt{2}}{2}(\vec{i}\,' - \vec{j}\,')dx + \frac{\sqrt{2}}{2}(\vec{i}\,' + \vec{j}\,')dy + \vec{k}\,dz \ , \qquad 4.10$$

and this relation matches the coordinate bases condition (*eq.4.1*) only by taking as basis vectors:

$$\frac{\sqrt{2}}{2}(\vec{i}\,' - \vec{j}\,') \ , \quad \frac{\sqrt{2}}{2}(\vec{i}\,' + \vec{j}\,') \ , \quad \vec{k}$$

that is nothing but the old coordinate basis $\{\vec{i}, \vec{j}, \vec{k}\}$. It goes without saying that the rotated basis $\{\vec{i}\,', \vec{j}\,', \vec{k}\}$ is not a coordinate basis (as we clearly see from *eq.4.10*).
This example shows that, as already said, *only in a coordinate basis* $d\vec{x} \overset{comp}{\rightarrow} dx^\mu$ is true. Not otherwise!

▪ The dual coordinate basis of covectors can be deduced from the duality relation $\langle \tilde{e}^\mu, \vec{e}_\nu \rangle = \delta^\nu_\mu$ in both cases, as follows:

❶ In 3D rectangular coordinates it is easy to see that the vector coordinate basis $\{\vec{e}_\nu\} \equiv \{\vec{i}, \vec{j}, \vec{k}\}$ overlaps the covector coordinate basis $\{\tilde{e}^\mu\} \equiv \{\tilde{i}, \tilde{j}, \tilde{k}\}$ obtained by duality *

---

* $\vec{i}$ and $\tilde{i}$ (in general $\vec{e}_\nu$ and $\tilde{e}^\mu$) remain distinct entities, although superimposable. In Cartesian they have the same expansion by components and their "arrows" coincide, but under change of coordinates they return to differ.

- Indeed, it is already known that in rectangular the coordinate basis is made by the unit vectors $\vec{i}, \vec{j}, \vec{k}$. They can be written in terms of components:

$$\vec{i} = (1,0,0) \quad , \quad \vec{j} = (0,1,0) \quad , \quad \vec{k} = (0,0,1)$$

Then the coordinate covector basis will necessarily be:

$$\tilde{i} = (1,0,0) \quad , \quad \tilde{j} = (0,1,0) \quad , \quad \tilde{k} = (0,0,1)$$

because in this way only you get:

$$\begin{array}{lll} \langle \tilde{i}, \vec{i} \rangle = 1 \;, & \langle \tilde{i}, \vec{j} \rangle = 0 \;, & \langle \tilde{i}, \vec{k} \rangle = 0 \;, \\ \langle \tilde{j}, \vec{i} \rangle = 0 \;, & \langle \tilde{j}, \vec{j} \rangle = 1 \;, & \langle \tilde{j}, \vec{k} \rangle = 0 \;, \\ \langle \tilde{k}, \vec{i} \rangle = 0 \;, & \langle \tilde{k}, \vec{j} \rangle = 0 \;, & \langle \tilde{k}, \vec{k} \rangle = 1 \;, \end{array}$$

as required by the condition of duality.

❷ In polar coordinates, on the contrary, the vector coordinate basis $\{\vec{e}_\nu\} \equiv \{\vec{e}_\rho, \vec{e}_\theta\}$ does not match with the covector basis deduced from duality condition $\{\tilde{e}^\nu\} \equiv \{\tilde{e}^\rho, \tilde{e}^\theta\}$.

- Indeed: unit vectors in polar coordinates are by definition: $\hat{e}_\rho = (1,0)$, $\hat{e}_\theta = (0,1)$. Thus, from *eq.4.9*:

$$\vec{e}_\rho = (1,0) \quad , \quad \vec{e}_\theta = (0,\rho)$$

Taking $\tilde{e}^\rho = (a,b)$, $\tilde{e}^\theta = (c,d)$, from $\langle \tilde{e}^\mu, \vec{e}_\nu \rangle = \delta^\mu_\nu \Rightarrow$

$$\Rightarrow \begin{cases} \left.\begin{array}{l} 1 = \langle \tilde{e}^\rho, \vec{e}_\rho \rangle = (a,b)\cdot(1,0) = a \\ 0 = \langle \tilde{e}^\rho, \vec{e}_\theta \rangle = (a,b)\cdot(0,\rho) = b\rho \Rightarrow b=0 \end{array}\right\} \Rightarrow \tilde{e}^\rho = (1,0) \\ \left.\begin{array}{l} 0 = \langle \tilde{e}^\theta, \vec{e}_\rho \rangle = (c,d)\cdot(1,0) = c \\ 1 = \langle \tilde{e}^\theta, \vec{e}_\theta \rangle = (c,d)\cdot(0,\rho) = d\rho \Rightarrow d = \dfrac{1}{\rho} \end{array}\right\} \Rightarrow \tilde{e}^\theta = (0,\dfrac{1}{\rho}) \end{cases}$$

Note that the dual switch tensor $\mathbf{G}$, even if previously defined, has no role in the calculation of a basis from its dual.

## 4.5 Change of the coordinate system

The change of coordinates, by itself, does not request the change of bases, but we must do it if we want to continue working in coordinate basis in the new system as well. As usual, the change of basis will be performed by means of a transformation matrix $\Lambda = [\Lambda^{\nu'}_\mu]$ which depends upon both old $\{x^\mu\}$ and new $\{x^{\nu'}\}$ coordinate systems (an apex $'$ marks the new one). But which $\Lambda$ must we use for a given transformation of coordinates?

To find the actual form of $\Lambda$ we impose to vector basis a twofold condition: that the starting basis is a coordinate basis (first equality) and that the arrival basis is a coordinated basis too (second equality):

$$d\vec{x} = \vec{e}_\mu d x^\mu = \vec{e}_{\nu'} d x^{\nu'} \qquad 4.11$$

Each new coordinate $x^{\nu'}$ will be related to all the old coordinates $x^1, x^2, \dots x^n$ by a function $x^{\nu'} = x^{\nu'}(x^1, x^2, \dots x^n)$ , i.e.

$$x^{\nu'} = x^{\nu'}(\dots, x^\mu, \dots)$$

whose total differential is:

$$dx^{\nu'} = \frac{\partial x^{\nu'}}{\partial x^\mu} dx^\mu$$

Substituting into the double equality *eq.4.11*:

$$d\vec{x} = \vec{e}_\mu dx^\mu = \vec{e}_{\nu'} \frac{\partial x^{\nu'}}{\partial x^\mu} dx^\mu \quad \Rightarrow \quad \vec{e}_\mu = \vec{e}_{\nu'} \frac{\partial x^{\nu'}}{\partial x^\mu} \qquad 4.12$$

A comparison with the change of basis-vectors *eq.3.3* (inverse, *from new to old*):

$$\vec{e}_\mu = \Lambda_\mu^{\nu'} \vec{e}_{\nu'}$$

leads to the conclusion that:

$$\Lambda_\mu^{\nu'} = \frac{\partial x^{\nu'}}{\partial x^\mu} \qquad 4.13$$

$\Rightarrow \Lambda_\mu^{\nu'}$ is the matrix element of $\Lambda$ complying with the definition *eq.3.7* that we were looking for.*

Since the *Jacobian matrix* of the transformation is defined as:

$$J \stackrel{\text{def}}{=} \begin{bmatrix} \frac{\partial x^{1'}}{\partial x^1} & \frac{\partial x^{1'}}{\partial x^2} & \cdots & \frac{\partial x^{1'}}{\partial x^n} \\ \frac{\partial x^{2'}}{\partial x^1} & \frac{\partial x^{2'}}{\partial x^2} & & \frac{\partial x^{2'}}{\partial x^n} \\ \vdots & & & \vdots \\ \frac{\partial x^{n'}}{\partial x^1} & \frac{\partial x^{n'}}{\partial x^2} & \cdots & \frac{\partial x^{n'}}{\partial x^n} \end{bmatrix} = \left[ \frac{\partial x^{\nu'}}{\partial x^\mu} \right] \qquad 4.14$$

* Or equally, as in the case, element of its transpose $\Lambda^T$.

we can identify:

$$\Lambda \equiv J \qquad 4.15$$

▪ The coordinate transformation has thus as *related matrix* the Jacobian matrix $J$ (whose elements are the partial derivatives of the new coordinates with respect to the old ones). It states, together with its inverse and transpose, how the old coordinate bases transform into the new coordinate bases, and consequently how the components of vectors and tensors transform when coordinates change.

> After a coordinate transformation we have to transform bases and components by means of the related matrix $\Lambda \equiv J$ (and its inverse / transpose) to continue working in a coordinate basis.

▪ As always, to avoid confusion, it is more practical to start from the transformation formulas and take care to balance the indexes; the indexes of $\Lambda$ will suggest the proper partial derivatives to be taken.

For example, $P_{\nu'} = \Lambda^{\mu}_{\nu'} P_{\mu}$ $\Rightarrow$ use the matrix $\Lambda^{\mu}_{\nu'} = \dfrac{\partial x^{\mu}}{\partial x^{\nu'}}$ .

Note that the upper / lower position of the index marked $'$ in partial derivatives agrees with the upper / lower position on the symbol $\Lambda$.

## 4.6 Contravariant and covariant tensors

We summarize the transformation laws for *components* of vectors and covectors *in a coordinate basis*:

- vector components: $V^{\nu'} = \dfrac{\partial x^{\nu'}}{\partial x^{\mu}} V^{\mu}$ (*contravariant* scheme)
- covector components: $P_{\nu'} = \dfrac{\partial x^{\mu}}{\partial x^{\nu'}} P_{\mu}$ (*covariant* scheme)

According to the traditional formulation of Tensor Calculus "by components", the two transformation schemes above are assumed as definitions of *contravariant vector* and *covariant vector* respectively.

The equivalence with the terminology used so far is:

- vector $\leftrightarrow$ contravariant vector
- covector (or "1-form") $\leftrightarrow$ covariant vector

- For a tensor of higher rank, for example $\binom{1}{2}$, the following law of transformation of the components applies:

$$T^{\alpha'}_{\mu'\nu'} = \frac{\partial x^{\alpha'}}{\partial x^{\alpha}} \frac{\partial x^{\mu}}{\partial x^{\mu'}} \frac{\partial x^{\nu}}{\partial x^{\nu'}} \, T^{\alpha}_{\mu\nu} \qquad 4.16$$

which is only a different way to write *eq.3.10*, in coordinate basis.
The generalization of *eq.3.10* to tensors of any rank is obvious.
In general, a tensor of rank $\binom{h}{k}$ is told contravariant of order (= rank) $h$ and covariant of order $k$.
Traditional texts begin here, using the transformation laws under coordinate transformation of components* as a definition: a tensor is defined as a multidimensional entity whose components transform according to example *eq.4.16*.

## 4.7 Affine tensors

As required by the traditional definition, tensors are those quantities that transform according to the contravariant or covariant schemes under *any* admissible coordinate transformation. If we restrict the class of the allowed coordinate transformations, increases the set of quantities that comply the schemes of "tensorial" (contravariant / covariant) transformation. In particular, if we restrict to consider *linear transformations* only, we identify the class of *affine* tensors, wider than that of tensors without further qualification.
The coordinate transformations to which we restrict in this case have the form $x^{\nu'} = a^{\nu'}_{\mu} x^{\mu}$ where $a^{\nu'}_{\mu}$ are purely numerical coefficients.

These coefficients can be arranged in a matrix $\left[a^{\nu'}_{\mu}\right]$ that expresses the law of coordinate transformation. But in this case also the transformation matrix for vector components, i.e. the Jacobian matrix $J$ given by $\Lambda^{\nu'}_{\mu} = \frac{\partial x^{\nu'}}{\partial x^{\mu}}$ is made by the same numerical coefficients because here is $\frac{\partial x^{\nu'}}{\partial x^{\mu}} = a^{\nu'}_{\mu}$ .

Hence, for affine tensors, the law of coordinate transformation and the transformation law for vector components coincide and are both expressed by the same matrix $\Lambda$ (the components of covectors transform as usual by the transposed inverse of $\Lambda$ ).

---

* Although the traditional "by components" approach avoids speaking about bases, it is implied that there *one always works in a coordinate basis* (both old and new).

## 4.8 Cartesian tensors

If we further restrict to linear *orthogonal* transformations, we identify the class of Cartesian tensors (wider than affine tensors' one).
A linear orthogonal transformation is represented by an orthogonal matrix.* Also for Cartesian tensors the same matrix $\left[a^{\nu'}_{\mu}\right]$ rules both the coordinate transformation and the transformation of vector components. Moreover, since an orthogonal matrix coincides with its transposed inverse, the same matrix $\left[a^{\nu'}_{\mu}\right]$ transforms as well covector components. Hence, since vectors and covectors transform in the same way, they are the same thing: the distinction between vectors and covectors (or contravariant tensors and covariant tensors) falls when it comes to Cartesian tensors.

## 4.9 Magnitude of vectors

We define *magnitude* (or *norm)* of a vector $\vec{V}$ the scalar $|\vec{V}|$ such that:

$$|\vec{V}|^2 = \vec{V} \bullet \vec{V} = \mathbf{G}(\vec{V}, \vec{V}) \qquad 4.17$$

Note that this definition, given by means of a scalar product, depends on the switch $\mathbf{G}$ that has been chosen.

## 4.10 Distance and metric tensor

In a manifold with a coordinate system $x^{\mu}$ a *distance ds* from a point *P* to a point *P'* shifted by $d\vec{x}$ can be defined as:

$$ds^2 = |d\vec{x}|^2 = \mathbf{G}(d\vec{x}, d\vec{x}) \qquad 4.18$$

In fact, what has been defined is an infinitesimal distance, also called "arc element".
After defined a distance the manifold becomes a ***metric space***.
Since its definition is given by scalar product, the distance depends on the $\mathbf{G}$ we use. If we want the distance to be defined in a certain way, it follows that $\mathbf{G}$ must be chosen appropriately.

We will denote $\mathbf{g}$ the tensor $\binom{0}{2}$ we pick out among the possible switches $\mathbf{G}$ to get the desired definition of distance.

---

* A matrix is orthogonal when its rows are orthogonal vectors (i.e. scalar product in pairs = 0), and so its columns too.

▫ Of course, definitions and properties of **G** now move onto **g** ; namely:

$\mathbf{g} \equiv$ tensor scalar product: $\vec{A} \bullet \vec{B} = \mathbf{g}(\vec{A}, \vec{B})$ *(eq.2.36)*

$g_{\alpha\beta} = \mathbf{g}(\vec{e}_\alpha, \vec{e}_\beta) = \vec{e}_\alpha \bullet \vec{e}_\beta$ *(eq.2.39)*

The distance *ds* given by the tensor equation:

$$ds^2 = \mathbf{g}(d\vec{x}, d\vec{x}) \qquad 4.19$$

is an invariant scalar and does not depend on the coordinate system. At this point, the tensor **g** defines the geometry of the manifold, i.e. its metric properties, based on the definition of distance between two points. For this reason **g** is called ***metric tensor***, or "metric".

▪ If (and only if) we use *coordinate bases*, the distance can be written:

$$ds^2 = g_{\alpha\beta}\, dx^\alpha dx^\beta \qquad 4.20$$

▫ Indeed, (only) in a coordinate basis $d\vec{x} = \vec{e}_\mu dx^\mu$ holds, hence:

$$ds^2 = \mathbf{g}(d\vec{x}, d\vec{x}) = \mathbf{g}(\vec{e}_\alpha dx^\alpha, \vec{e}_\beta dx^\beta) = \mathbf{g}(\vec{e}_\alpha, \vec{e}_\beta)\, dx^\alpha dx^\beta =$$
$$= g_{\alpha\beta}\, dx^\alpha dx^\beta$$

## 4.11 Euclidean distance

If the manifold is the usual Euclidean space $\mathbb{R}^3$ , using a Cartesian rectangular coordinate system, the distance is classically defined:

$$ds = \sqrt{dx^2 + dy^2 + dz^2} \qquad 4.21$$

or:

$$ds^2 = (dx^1)^2 + (dx^2)^2 + (dx^3)^2 \qquad 4.22$$

which express the Pythagoras theorem and coincides with the form *(eq.4.20)* that applies in coordinate bases

$$ds^2 = (dx^1)^2 + (dx^2)^2 + (dx^3)^2 = g_{\alpha\beta}\, dx^\alpha dx^\beta \qquad 4.23$$

provided we take:

$$g_{\alpha\beta} = \begin{cases} 1 \text{ for } \alpha = \beta \\ 0 \text{ for } \alpha \neq \beta \end{cases} \qquad \text{i.e.} \qquad \mathbf{g} = \begin{bmatrix} 1 & 0 & 0 \\ 0 & 1 & 0 \\ 0 & 0 & 1 \end{bmatrix} = diag(+1) \qquad 4.24$$

Note that in this case $\mathbf{g} = \mathbf{g}^{-1}$, namely $g_{\alpha\beta} = g^{\alpha\beta}$ .

## 4.12 Generalized distances

The definition of distance given by *eq.4.20* is more general than the Euclidean case and includes any bilinear form in $dx^1, dx^2, \ldots dx^n$ such as:

$$g_{11}dx^1dx^1+g_{12}dx^1dx^2+g_{13}dx^1dx^3+\ldots g_{21}dx^2dx^1+g_{22}dx^2dx^2+\ldots g_{nn}dx^ndx^n \qquad 4.25$$

where $g_{\alpha\beta}$ are subject to few restrictions. Inasmuch they build up the matrix that represents the metric tensor **g**, we must at least require that their matrix is symmetric and $det[g_{\alpha\beta}]\neq 0$ so that **g** is invertible in order $\mathbf{g}^{-1}$ to exist.

▫ Note, however, that in any case **g** is symmetrisable: for example, if it were $g_{12}\neq g_{21}$ in the bilinear form *eq.4.25*, just replacing them by $g'_{12}=g'_{21}=(g_{12}+g_{21})/2$ nothing would change in the definition of distance.

Metric spaces where the distance is defined by *eq.4.19* with any **g**, provided symmetric and invertible, are called ***Riemann spaces***.

▪ Given a manifold, its metric properties are fully described by the metric tensor **g** associated to it. As tensor, **g** does not depend on the coordinate system imposed onto the manifold: **g** does not change if the coordinate system changes. However, we do not know how to represent **g** by itself: its representation is only possible in terms of components $g_{\alpha\beta}$ and they do depend upon the particular coordinate system (and therefore upon coordinate basis). For a given **g** we then have different matrices $[g_{\alpha\beta}]$, one for each coordinate system, which mean the same **g**, i.e. the same metric properties for the manifold.

For instance, for the 2D Euclidean plane, the metric tensor **g**, which expresses the usual Euclidean metric properties is represented by $[g_{\alpha\beta}]=\begin{bmatrix}1 & 0\\ 0 & 1\end{bmatrix}$ in Cartesian coordinates and by $[g_{\alpha'\beta'}]=\begin{bmatrix}1 & 0\\ 0 & \rho^2\end{bmatrix}$ in polar coordinates. Of course, there are countless other $[g_{\mu'\nu'}]$ which can be obtained from the previous ones by coordinate transformation and consequent change of basis by means of the related matrix $\Lambda$ (remind that $[g_{\alpha\beta}]\overset{\Lambda^{\alpha}_{\mu'}\Lambda^{\beta}_{\nu'}}{\rightarrow}[g_{\mu'\nu'}]$, i.e. $g_{\mu'\nu'}=\Lambda^{\alpha}_{\mu'}\Lambda^{\beta}_{\nu'}\,g_{\alpha\beta}$ ).

In other words, the same Euclidean metric can be expressed by all the

$[g_{\mu'\nu'}]$ attainable by coordinate transformation from the Cartesian rectangular $[g_{\alpha\beta}] = \begin{bmatrix} 1 & 0 \\ 0 & 1 \end{bmatrix}$ .

▪ Summary: $\left.\begin{array}{l} \text{geometry of manifold} \Leftrightarrow \mathbf{g} \Rightarrow \\ \text{coordinate system} \quad\quad\quad \Rightarrow \end{array}\right\} \Rightarrow g_{\mu\nu}$

## 4.13 Tensors and not -

The contra / covariant transformation laws (exemplified by *eq.4.16*) provide a criterion to state whether a mathematical quantity is a tensor:

> only if its components change according to the schemes contra / covariant (*eq.4.16*), the quantity is a tensor.

For example, under coordinate transformation $x^{\nu} \rightarrow x^{\alpha'}$ the components of the infinitesimal displacement $d\vec{x}$ transform:

$$dx^{\nu'} = \frac{\partial x^{\nu'}}{\partial x^{\mu}} dx^{\mu}$$

(≡ total differential of the function $x^{\alpha'} = x^{\alpha'}(x^{\nu})$ ) confirming that the displacement vector $d\vec{x}$ is a contravariant tensor of the first order.

▪ Instead, is not a tensor the “position vector” $\vec{x}$ whose components transform according to the law of coordinate transformation (which in general has nothing to do with the contra / covariant schemes).

▪ The derivative of vector $\frac{\partial \vec{V}}{\partial x^{\mu}}$ is a tensor too; but its components are *not* simply partial derivatives. In fact:
Taken a vector $\vec{V} = \vec{e}_{\nu} V^{\nu}$ let us build a vector $\vec{V}\,'$ whose components are the derivatives of the components of $\vec{V}$ ; its expansion is:

$$\vec{V}\,' = \vec{e}_{\nu} \frac{\partial V^{\nu}}{\partial x^{\mu}} \quad \text{or, in short form} \quad \vec{V}\,' = \vec{e}_{\nu}\, \partial_{\mu} V^{\nu}$$

Let's check now whether it is a tensor.
From (inverse) transformation of components of the vector $\vec{V}$ :

$$V^{\nu} = V^{\alpha'} \frac{\partial x^{\nu}}{\partial x^{\alpha'}} \qquad 4.26$$

we get, taking its derivative, the transformation rule for derivatives of the components of $\vec{V}'$ : *

$$\frac{\partial V^{\nu}}{\partial x^{\mu}} = \frac{\partial V^{\alpha'}}{\partial x^{\mu}} \frac{\partial x^{\nu}}{\partial x^{\alpha'}} + V^{\alpha'} \frac{\partial}{\partial x^{\mu}} \frac{\partial x^{\nu}}{\partial x^{\alpha'}} = \frac{\partial V^{\alpha'}}{\partial x^{\kappa'}} \frac{\partial x^{\kappa'}}{\partial x^{\mu}} \frac{\partial x^{\nu}}{\partial x^{\alpha'}} + V^{\alpha'} \frac{\partial^2 x^{\nu}}{\partial x^{\mu} \partial x^{\alpha'}} \qquad 4.27$$

that is *not* the tensor components' transformation scheme *eq.4.16* because of the presence of an additional $\partial^2$ term. It follows that $\vec{V}'$ is *not* a tensor. However, against the first appearance, $\vec{V}'$ *is not the derivative of vector* $\vec{V}$ .
In the next paragraph a true vector "derivative of vector" $\partial_{\mu}\vec{V}$ will be built and it will appear to be a tensor.

## 4.14 Covariant derivative

▪ Indeed, the derivative of a vector (itself a vector) does not end just in the derivatives of the components, because even the basis-vectors vary from point to point in the manifold: the derivative deals with basis vectors too:

$$\frac{\partial \vec{V}}{\partial x^{\mu}} \equiv \partial_{\mu}\vec{V} = \partial_{\mu}(\vec{e}_{\nu} V^{\nu}) = \vec{e}_{\nu}\partial_{\mu}V^{\nu} + V^{\nu}\partial_{\mu}\vec{e}_{\nu} \qquad 4.28$$

Of the two terms, the first describes the variability of the components of the vector, the second the variability of the basis-vectors along the coordinate lines $x^{\mu}$ .
We observe that $\partial_{\mu}\vec{e}_{\nu}$, the partial derivative of a basis-vector, although *not* a tensor (see *Appendix*) is a vector in a geometric sense (because it represents the change of the basis vector $\vec{e}_{\nu}$ along the infinitesimal arc $dx^{\mu}$ on the coordinate line $x^{\mu}$ ) and therefore can be expanded on the basis of vectors.
Hence, setting $\partial_{\mu}\vec{e}_{\nu} = \vec{\Gamma}$ , we can expand it:

$$\vec{\Gamma} = \Gamma^{\lambda}\vec{e}_{\lambda} \qquad 4.29$$

But in reality $\vec{\Gamma}$ is already labeled with lower indexes $\mu, \nu$ , hence $\vec{\Gamma}_{\mu\nu} = \Gamma^{\lambda}_{\mu\nu}\vec{e}_{\lambda}$ , so that:

$$\partial_{\mu}\vec{e}_{\nu} = \Gamma^{\lambda}_{\mu\nu}\vec{e}_{\lambda} \qquad 4.30$$

The vectors $\vec{\Gamma}_{\mu\nu}$ are $n^2$ in number, one for each pair $\mu, \nu$ , with $n$

---

* Leibniz rule for the derivative of product: $(f \cdot g)' = f' \cdot g + f \cdot g'$ and then the "chain rule".

components $\Gamma^{\lambda}_{\mu\nu}$ each (for $\lambda = 1,2,\dots n$).

The coefficients $\Gamma^{\lambda}_{\mu\nu}$ are called ***Christoffel symbols*** or ***connection coefficients***; they are in total $n^3$ coefficients.

*Eq.4.30* is equivalent to:

$$\partial_\mu \vec{e}_\nu \overset{comp}{\rightarrow} \Gamma^{\lambda}_{\mu\nu} \qquad 4.31$$

> The Christoffel symbols $\Gamma^{\nu}_{\mu\lambda}$ are therefore the components of the vectors "partial derivative of a basis vector" on the basis of the vectors themselves
>
> *or*:
>
> The "recipe" of the partial derivative of a basis vector uses as "ingredients" the same basis vectors while Christoffel symbols represent their amounts.

Inserting *eq.4.30* into *eq.4.28,* the latter becomes:

$$\partial_\mu \vec{V} = \vec{e}_\nu \partial_\mu V^\nu + \Gamma^{\lambda}_{\mu\nu} \vec{e}_\lambda V^\nu =$$

in the last right term $\nu, \lambda$ are dummy summation indexes: they can therefore be freely changed (and interchanged). Interchanging $\nu, \lambda$ :

$$= \vec{e}_\nu \partial_\mu V^\nu + \Gamma^{\nu}_{\mu\lambda} \vec{e}_\nu V^\lambda = \vec{e}_\nu \left( \partial_\mu V^\nu + \Gamma^{\nu}_{\mu\lambda} V^\lambda \right)$$

In short:

$$\partial_\mu \vec{V} = \vec{e}_\nu \underbrace{\left( \partial_\mu V^\nu + \Gamma^{\nu}_{\mu\lambda} V^\lambda \right)}_{V^{\nu}_{;\mu}} \qquad 4.32$$

We define ***covariant derivative of a vector*** with respect to $x^\mu$ :

$$V^{\nu}_{;\mu} \overset{\text{def}}{=} \partial_\mu V^\nu + \Gamma^{\nu}_{\mu\lambda} V^\lambda \qquad 4.33$$

(we introduce here the subscript ***;*** to denote the *covariant derivative*).

The covariant derivative $V^{\nu}_{;\mu}$ is a scalar which differs from the respective partial derivative $\partial_\mu V^\nu$ by a corrective term in $\Gamma$ .

So far, we did nothing but writing in a new form the derivative of a vector, introducing by the way the Christoffel symbol $\Gamma$ and the covariant derivative.

▪ The important fact is that, unlike the ordinary derivative, *the covariant derivative has tensorial features* in the sense that the $n^2$ covariant derivatives *eq.4.33* transform according to *eq.4.16* (see in

*Appendix* a check *in extenso*) and can therefore be considered components of tensors (beginning from the vector derivative $\partial_\mu \vec{V}$ ).

▪ From *eq.4.32* we see that the derivative of a vector is a vector whose components are the covariant derivatives of the vector, that means a tensor:

$$\partial_\mu \vec{V} = \vec{e}_\nu V^\nu_{\ ;\mu} \qquad 4.34$$

> The partial derivative of a vector is a vector (tensor) whose scalar components are the *n* covariant derivatives of the vector.

▪ When covariant derivative and partial derivative identify?
From *eq.4.33* we see that for this to happen it must be $\forall \Gamma = 0$. Since *eq.4.30* means the implication

$$\forall \vec{e}_\nu = constant \quad \Leftrightarrow \quad \forall \Gamma = 0$$

we deduce that:

> when all basis-vectors $\vec{e}_\nu$ are constant the covariant derivative equals the partial derivative. This is the case of Euclidean-Cartesian spaces.

It is in these same cases that the derivative of a vector takes the simple form:

$$\frac{\partial \vec{V}}{\partial x^\mu} = \vec{e}_\nu \frac{\partial V^\nu}{\partial x^\mu}$$

▪ In a similar way the case of derivative of covector is dealt:

$$\frac{\partial \tilde{A}}{\partial x^\mu} = \partial_\mu \tilde{A} = \partial_\mu (\tilde{e}^\nu A_\nu) = \tilde{e}^\nu \partial_\mu A_\nu + A_\nu \partial_\mu \tilde{e}^\nu \qquad 4.35$$

As in the previous case, the term $\partial_\mu \tilde{e}^\nu$ describes the variability of the basis-covector $\tilde{e}^\nu$ along the coordinate line $x^\mu$ and, though not a covector in tensorial sense, can be expanded on the basis of covectors.

Set $\partial_\mu \tilde{e}^\nu = \tilde{L}$ , we write:

$$\tilde{L} = L_\lambda \tilde{e}^\lambda$$

Really, $\tilde{L}$ is labeled with an upper $\nu$ and a lower $\mu$ index, then:

$$\tilde{L}^\nu_\mu = L^{\ \nu}_{\mu\lambda} \tilde{e}^\lambda$$

hence:

$$\partial_\mu \tilde{e}^\nu = L^{\ \nu}_{\mu\lambda} \tilde{e}^\lambda \qquad 4.36$$

which, placed into *eq.4.35*, gives:

$$\partial_\mu \tilde{A} = \tilde{e}^\nu \partial_\mu A_\nu + L^\nu_{\mu\lambda} \tilde{e}^\lambda A_\nu =$$

interchanging the dummy indexes $\nu, \lambda$ in the last term:

$$= \tilde{e}^\nu \partial_\mu A_\nu + L^\lambda_{\mu\nu} \tilde{e}^\nu A_\lambda = \tilde{e}^\nu (\partial_\mu A_\nu + L^\lambda_{\mu\nu} A_\lambda) \qquad 4.37$$

▪ How are related $\Gamma$ and L ? They simply differ by a sign:

$$L^\lambda_{\mu\nu} = -\Gamma^\lambda_{\mu\nu} \qquad 4.38$$

▫ It can be shown by calculating separately the derivatives of both right and left members of the duality condition $\langle \tilde{e}^\mu, \vec{e}_\nu \rangle = \delta^\mu_\nu$ . Since $\delta^\mu_\nu = 0$ or $1$ (however a constant), then $\partial_\kappa(\delta^\mu_\nu) = 0$ .

On the other hand: $\partial_\kappa \langle \tilde{e}^\mu, \vec{e}_\nu \rangle = \langle \partial_\kappa \tilde{e}^\mu, \vec{e}_\nu \rangle + \langle \tilde{e}^\mu, \partial_\kappa \vec{e}_\nu \rangle$

$$= \langle L^\mu_{\kappa\lambda} \tilde{e}^\lambda, \vec{e}_\nu \rangle + \langle \tilde{e}^\mu, \Gamma^\lambda_{\kappa\nu} \vec{e}_\lambda \rangle$$

$$= L^\mu_{\kappa\lambda} \langle \tilde{e}^\lambda, \vec{e}_\nu \rangle + \Gamma^\lambda_{\kappa\nu} \langle \tilde{e}^\mu, \vec{e}_\lambda \rangle$$

$$= L^\mu_{\kappa\lambda} \delta^\lambda_\nu + \Gamma^\lambda_{\kappa\nu} \delta^\mu_\lambda$$

$$= L^\mu_{\kappa\nu} + \Gamma^\mu_{\kappa\nu}$$

Hence $0 = L^\mu_{\kappa\nu} + \Gamma^\mu_{\kappa\nu} \Rightarrow$ *eq.4.38*, q.e.d.

Therefore: $$\partial_\mu \tilde{e}^\nu = -\Gamma^\nu_{\mu\lambda} \tilde{e}^\lambda \qquad 4.39$$

and from *eq.4.37*: $$\partial_\mu \tilde{A} = \tilde{e}^\nu \underbrace{(\partial_\mu A_\nu - \Gamma^\lambda_{\mu\nu} A_\lambda)}_{A_{\nu;\mu}} \qquad 4.40$$

We define ***covariant derivative of a covector*** with respect to $x^\mu$ :

$$A_{\nu;\mu} \stackrel{\text{def}}{=} \partial_\mu A_\nu - \Gamma^\lambda_{\mu\nu} A_\lambda \qquad 4.41$$

analogous to *eq.4.33*.

▪ The derivative of a covector can be written (from *eq.4.40*):

$$\partial_\mu \tilde{A} = \tilde{e}^\nu A_{\nu;\mu} \qquad 4.42$$

> The partial derivative of a covector is a covector whose scalar components are *n* covariant derivatives of the covector.

## 4.15 The gradient $\tilde{\nabla}$ at work

We have already stated that the gradient $\tilde{\nabla}$ is a covector whose components *in coordinate bases* are the partial derivatives with respect to the coordinates (*eq.4.4*, *eq.4.5*):

$$\tilde{grad} \equiv \tilde{\nabla} \overset{comp}{\rightarrow} \partial_\mu$$

namely

$$\tilde{\nabla} = \tilde{e}^\mu \partial_\mu \qquad 4.43$$

- That $\tilde{\nabla}$ is a tensor was deduced from *eq.4.3* in case of $\tilde{\nabla}$ applied to $f$, but it is in general true, confirmed by the fact that the "chain rule" applied to its components $\partial_{\nu'} = \partial_\mu \dfrac{\partial x^\mu}{\partial x^{\nu'}}$ coincides with the covariant transformation scheme *eq.4.16*.

Inasmuch a tensor, several tensor operations can be performed by $\tilde{\nabla}$ applying it to other tensors. We will examine the cases:

- outer product with vector, covector or tensor ⇒ ***gradient***
- inner product with vector or tensor ⇒ ***divergence***

### ❶ Tensor outer product $\tilde{\nabla} \otimes$ vector

The result is the tensor $\binom{1}{1}$ ***gradient of vector***. *In coordinate bases*:

$$\tilde{\nabla} \otimes \vec{V} = \tilde{e}^\mu \partial_\mu \otimes \vec{V}$$

that is to say: *

$$\tilde{\nabla} \vec{V} = \tilde{e}^\mu \partial_\mu \vec{V} \qquad 4.44$$

$\partial_\mu \vec{V}$ comes from *eq.4.32* ; substituting:

$$\tilde{\nabla} \vec{V} = \tilde{e}^\mu \vec{e}_\nu (\partial_\mu V^\nu + \Gamma^\nu_{\mu\lambda} V^\lambda) \qquad 4.45$$

On the other hand, $\tilde{\nabla} \vec{V}$ may be formally written as expansion on its components we'll denote by $\nabla_\mu V^\nu$ :

$$\tilde{\nabla} \vec{V} = \nabla_\mu V^\nu \tilde{e}^\mu \vec{e}_\nu \qquad 4.46$$

and by comparison between the last two we see that

$$\nabla_\mu V^\nu = \partial_\mu V^\nu + \Gamma^\nu_{\mu\lambda} V^\lambda \qquad 4.47$$

---

* It is convenient to think of the symbol ⊗ as a connective between tensors. In T-mosaic it means to juxtapose and past the two blocks $\tilde{e}^\mu \partial_\mu$ and $\vec{V}$

or, in the alternative notation suggested by *eq.4.33:*

$$\nabla_\mu V^\nu \equiv V^\nu_{;\mu} \qquad 4.48$$

*Eq.4.47* qualifies the *covariant derivative of vector* as component of the tensor $\tilde{\nabla}\vec{V}$ $(\equiv \tilde{\nabla} \otimes \vec{V})$ gradient of vector, while introducing a further notation $\nabla_\mu$ for the covariant derivative.

> The covariant derivatives of a vector $\nabla_\mu V^\nu$ are the $n^2$ components of tensor "gradient of vector" $\tilde{\nabla} \otimes \vec{V}$ (usually written $\tilde{\nabla}\vec{V}$ ).

- In *Appendix* it is shown that the covariant derivatives $V^\nu_{;\mu}$ transform according to the tensor contra / covariant scheme exemplified by *eq.4.16* and this confirms that the gradient $\tilde{\nabla}\vec{V}$ of the vector of which they are components is indeed a tensor.

▪ The partial derivative of a vector (*eq.4.34*) may be rewritten in the new notation in terms of $\tilde{\nabla}$ as:

$$\partial_\mu \vec{V} = \vec{e}_\nu \nabla_\mu V^\nu \qquad 4.49$$

that is:

$$\partial_\mu \vec{V} \overset{comp}{\rightarrow} \nabla_\mu V^\nu \qquad 4.50$$

▪ What we have seen about gradient tensor and derivative of a vector can be summarized in the sketch:

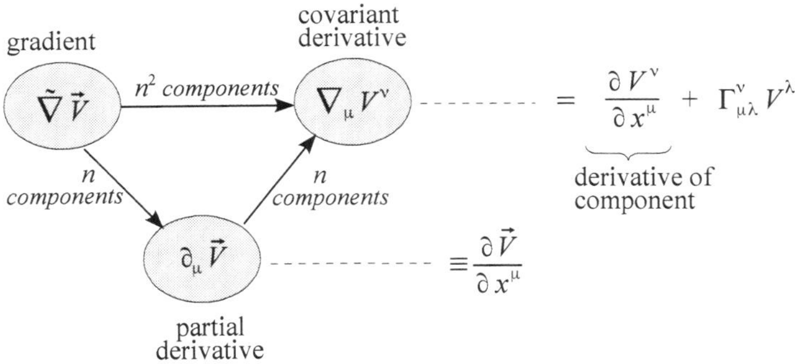


compliant with *eq.4.44*, *eq.4.49*:

$$\tilde{\nabla}\vec{V} = \tilde{e}^\mu \partial_\mu \vec{V} = \tilde{e}^\mu \vec{e}_\nu \nabla_\mu V^\nu$$

- In particular, when the vector is a basis-vector $\vec{e}_\nu$ the sketch above reduces to:

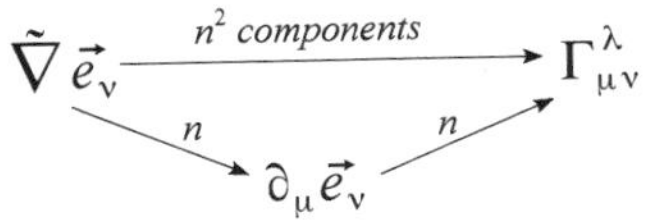


according to *eq.4.44*, *eq.4.31*:

$$\tilde{\nabla}\vec{e}_\nu = \tilde{e}^\mu \ \partial_\mu \vec{e}_\nu = \tilde{e}^\mu \vec{e}_\lambda \Gamma^\lambda_{\mu\nu} \ . \qquad 4.51$$

*Terminology*

We reserve the term "covariant derivative" to the scalar components which all together form the "gradient tensor":

$$\underbrace{\tilde{\nabla}\vec{V}}_{\text{gradient tensor}} \ \overset{comp}{\rightarrow} \ \underbrace{\nabla_\mu V^\nu}_{\text{covariant derivative}} \ = \ \underbrace{\partial_\mu V^\nu}_{\text{partial derivative}} \ + \ \underbrace{\Gamma^\nu_{\mu\lambda}}_{\text{Christoffel symbol}} \ V^\lambda$$

*Notation*

The covariant derivative of $V^\nu$ with respect to $x^\mu$ is denoted by

$$\nabla_\mu V^\nu \qquad \text{or} \qquad V^\nu_{\ ;\mu}$$

Not to be confused with the partial derivative $\dfrac{\partial V^\nu}{\partial x^\alpha}$ , denoted by

$$\partial_\mu V^\nu \qquad \text{or} \qquad V^\nu_{\ ,\mu}$$

The lower index after the , means partial derivative
The lower index after the ; means covariant derivative

### ❷ Tensor outer product $\tilde{\nabla}\otimes$ covector

The result is the $\binom{0}{2}$ tensor ***gradient of covector***. *In coordinate bases*:

$$\tilde{\nabla}\otimes\tilde{A} = \tilde{e}^\mu \partial_\mu \otimes \tilde{A}$$

or

$$\tilde{\nabla}\tilde{A} = \tilde{e}^\mu \partial_\mu \tilde{A} \qquad 4.52$$

We go on as in the case of gradient of a vector:
$\partial_\mu \tilde{A}$ comes from *eq.4.40*; replacing it:

$$\tilde{\nabla}\tilde{A} = \tilde{e}^\mu \tilde{e}^\nu (\partial_\mu A_\nu - \Gamma^\lambda_{\mu\nu} A_\lambda) \qquad 4.53$$

and by comparison with the generic definition

$$\tilde{\nabla}\tilde{A} = \nabla_\mu A_\nu \, \tilde{e}^\mu \tilde{e}^\nu \qquad 4.54$$

we get:

$$\nabla_\mu A_\nu = \partial_\mu A_\nu - \Gamma^\lambda_{\mu\nu} A_\lambda \qquad 4.55$$

which is equivalent to *eq.4.41* with the alternative notation:

$$\nabla_\mu A_\nu \equiv A_{\nu;\mu} \qquad 4.56$$

*Eq.4.55* represents the *covariant derivative of covector* and qualifies it as a component of the tensor gradient of covector $\tilde{\nabla}\tilde{A}$ $(\equiv \tilde{\nabla} \otimes \tilde{A})$. *Eq.4.55* is the analogous of *eq.4.47*, which was stated for vectors.

- The partial derivative of a covector (*eq.4.42*) may be written in the new notation in terms of $\tilde{\nabla}$ as:

$$\partial_\mu \tilde{A} = \vec{e}_\nu \nabla_\mu A_\nu \qquad 4.57$$

which means:

$$\partial_\mu \tilde{A} \overset{comp}{\rightarrow} \nabla_\mu A_\nu \qquad 4.58$$

- For the gradient and the derivative of a covector a sketch like that drawn for vectors applies, according to *eq.4.52*, eq.*4.57*:

$$\tilde{\nabla}\tilde{A} = \tilde{e}^\mu \partial_\mu \tilde{A} = \vec{e}_\nu \tilde{e}^\mu \nabla_\mu A_\nu$$

*Mnemo*

To write the covariant derivative of vectors and covectors:

- we had better writing the $\Gamma$ before the component, not vice versa
- 1st step:
  $\nabla_\mu A^\nu = \partial_\mu A^\nu + \Gamma^\nu_\mu$ or $\nabla_\mu A_\nu = \partial_\mu A_\nu - \Gamma_{\mu\nu}$
  (adjust the indexes of $\Gamma$ according to those of the 1st member)
- 2nd step: paste a dummy index to $\Gamma$ and to the component $A$ that follows in order to balance the upper / lower indexes:
  $\ldots\ldots\ldots + \Gamma^\nu_{\mu\kappa} A^\kappa$ or $\ldots\ldots\ldots - \Gamma^\kappa_{\mu\nu} A_\kappa$

## Tensor outer product $\tilde{\nabla} \otimes$ tensor $\binom{h}{k}$

$\tilde{\nabla} \otimes \mathbf{T}$ (or $\tilde{\nabla}\mathbf{T}$) is a $\binom{h}{k+1}$ tensor, the ***gradient of tensor.***

By the same procedure of previous cases we can find that, for example for a rank $\binom{2}{1}$ tensor $\mathbf{T} = T^{\alpha\beta}_\gamma \, \vec{e}_\alpha \vec{e}_\beta \tilde{e}^\gamma$ is:

$$\tilde{\nabla}\mathbf{T} = \nabla_\mu T^{\alpha\beta}_\gamma \vec{e}_\mu \vec{e}_\alpha \vec{e}_\beta \tilde{e}^\gamma \qquad \text{namely:}$$

$$\tilde{\nabla}\mathbf{T} \stackrel{comp}{\rightarrow} \nabla_\mu T^{\alpha\beta}_\gamma = \partial_\mu T^{\alpha\beta}_\gamma + \Gamma^\alpha_{\mu\kappa} T^{\kappa\beta}_\gamma + \Gamma^\beta_{\mu\kappa} T^{\alpha\kappa}_\gamma - \Gamma^\kappa_{\mu\gamma} T^{\alpha\beta}_\kappa \qquad 4.59$$

In general, the ***covariant derivative of a tensor*** $\binom{h}{k}$ is obtained by attaching to the partial derivative $h$ terms in $\Gamma$ with positive sign and $k$ terms in $\Gamma$ with negative sign. Treat the indexes individually, one after the other in succession.

Covariant derivatives are $n^{r+1}$ in number if the tensor is ranked $r$.
Each covariant derivative consists of $r$ terms in $\Gamma$ since each one takes into account the variability of a single index of basis-vector (o covector), whose total number is $r$ .

*Mnemo*

To write the covariant derivative of tensors, for example $\nabla_\mu T^{\alpha\beta}_\gamma$ :
- after the term in $\partial_\mu$ consider the indexes of $T$ one by one:
- write the first term in $\Gamma$ as it were $\nabla_\mu T^\alpha$, then complete $T$ with its remaining indexes ${}^\beta_\gamma$ : $\rightarrow +\Gamma^\alpha_{\mu\kappa} \rightarrow +\Gamma^\alpha_{\mu\kappa} T^{\kappa\beta}_\gamma$
- write the second term in $\Gamma$ as it were $\nabla_\mu T^\beta$, then complete $T$ with its remaining indexes ${}^\alpha_\gamma$ : $\rightarrow +\Gamma^\beta_{\mu\kappa} \rightarrow +\Gamma^\beta_{\mu\kappa} T^{\alpha\kappa}_\gamma$
- write the third $\Gamma$ *as it were* $\nabla_\mu T_\gamma$ , then complete $T$ with its remaining indexes ${}^{\alpha\beta}$ : $\rightarrow -\Gamma^\kappa_{\mu\gamma} \rightarrow -\Gamma^\kappa_{\mu\gamma} T^{\alpha\beta}_\kappa$

- **Considerations on the form of components of tensor $\tilde{\nabla}$**

$\tilde{\nabla} \stackrel{comp}{\rightarrow} \partial_\mu$ holds in general for tensors of any rank $r$ :

$$\tilde{\nabla}\mathbf{T} \stackrel{comp}{\rightarrow} \partial_\mu \mathbf{T}$$

$\partial_\mu \mathbf{T}$ are in any case the $n$ components of $\tilde{\nabla}\mathbf{T}$ ; however, except the case $r = 0$ , they are *not the scalar components but just the first level-components*, $n$ in number.
Instead, scalar components of $\tilde{\nabla}\mathbf{T}$ are the $n^{r+1}$ covariant derivatives $\nabla_\mu \ldots$ ($r$+1 is the rank of $\tilde{\nabla}\mathbf{T}$ when $\mathbf{T}$ has rank $r$ ).
For $r = 0$, or $\mathbf{T} \equiv f$ scalar, first level-components $\partial_\mu \mathbf{T}$ and scalar components coincide. For $r=1$, $\partial_\mu \mathbf{T}$ are vectors and for $r>1$ they are tensors.

We write in general:

$$\tilde{\nabla} \overset{comp}{\rightarrow} \nabla_{\mu} \qquad 4.60$$

where $\nabla_{\mu}$ are the covariant derivatives (scalar components).

For the very way in which they are defined, the covariant derivatives $\nabla_{\mu}$ automatically assume different form depending on the object to which the gradient $\tilde{\nabla}$ is applied:

- $\partial_{\mu}$ vhen $\tilde{\nabla}$ is applied to a scalar
- $\partial_{\mu} + \Gamma \dots$ when applied to a vector $\vec{V}$
- $\partial_{\mu} - \Gamma \dots$ when applied to a covector $\tilde{P}$
- $\partial_{\mu} \pm$ various terms in $\Gamma \dots$ according to the rank of tensor **T**

increasing in complexity and number as $r$ increases.

**❹ Inner product $\tilde{\nabla} \bullet$ vector**

$$\tilde{\nabla}(\vec{V}) = \nabla_{\mu} V^{\mu} = V^{\mu}_{;\mu} = div\, \vec{V} \qquad 4.61$$

It is the scalar ***divergence of a vector***.

Note that $\tilde{\nabla}$ applied as inner product to a vector looks like a covariant derivative:

$$\nabla_{\mu} V^{\mu} = \partial_{\mu} V^{\mu} + \Gamma^{\mu}_{\mu\kappa} V^{\kappa} \qquad 4.62$$

but with equal indexes (dummy), and is actually a number.

> The divergence of a vector $div\, \vec{V}$ looks like a covariant derivative, except for a single repeated index: $\nabla_{\mu} V^{\mu}$ or $V^{\mu}_{;\mu}$

To justify the presence of additional terms in $\Gamma$ it is appropriate to think of the divergence $\nabla_{\mu} V^{\mu}$ as a covariant derivative $\nabla_{\nu} V^{\mu}$ followed by an index contraction $\nu = \mu$ (strictly speaking it is a contraction of the gradient):

$$\tilde{\nabla} \otimes \vec{V} = \nabla_{\nu} V^{\mu}\, \tilde{e}^{\nu} \otimes \vec{e}_{\mu} \xrightarrow{\text{contraction } \nu=\mu} \nabla_{\mu} V^{\mu} \qquad 4.63$$

which results in a scalar product, that is a scalar.

This definition of divergence is a direct generalization of the divergence as known from Vector Analysis, where it is defined in Cartesian coordinates as the sum of the partial derivatives of components of the vector. In fact, in a Cartesian frame, it is $div\, \vec{V} = \nabla_{\mu} V^{\mu} = \partial_{\mu} V^{\mu}$ because the coefficients $\Gamma$ are null.

## ❺ Inner product $\tilde{\nabla}\cdot$ tensor $\binom{h}{k}$

Divergence(s) of tensors of higher order can similarly be defined provided at least one upper index exists.

It results in a tensor of rank $\binom{h-1}{k}$, tensor ***divergence of a tensor***.

If there are more than one upper index, various divergences can be defined, one for each *upper* index. For each index of the tensor, both upper or lower, including the one involved, an adequate term in $\Gamma$ must be added. For example, the divergence with respect to $\beta$ index of the rank $\binom{2}{1}$ tensor $\mathbf{T} = T^{\alpha\beta}_{\gamma}\,\vec{e}_\alpha\,\vec{e}_\beta\,\tilde{e}^\gamma$ is the $\binom{1}{1}$ tensor:

$$\tilde{\nabla}(\mathbf{T}) = \nabla_\beta T^{\alpha\beta}_{\gamma}\,\vec{e}_\alpha\,\tilde{e}^\gamma \qquad \text{namely:}$$

$$\tilde{\nabla}(\mathbf{T}) \overset{comp}{\rightarrow} \nabla_\beta T^{\alpha\beta}_{\gamma} = \partial_\beta T^{\alpha\beta}_{\gamma} + \Gamma^{\alpha}_{\beta\kappa} T^{\kappa\beta}_{\gamma} + \Gamma^{\beta}_{\beta\kappa} T^{\alpha\kappa}_{\gamma} - \Gamma^{\kappa}_{\beta\gamma} T^{\alpha\beta}_{\kappa} \qquad 4.64$$

Note that it is the contraction of $\tilde{\nabla}\mathbf{T}$ (*eq.4.59*) for $\mu = \beta$.

## 4.16 Gradient of some fundamental tensors

It is interesting to calculate the gradient of some fundamental tensors: the identity tensor and the metric tensor.

- $\tilde{\nabla}\mathbf{I} \overset{comp}{\rightarrow} \nabla_\mu \delta^\alpha_\beta = \partial_\mu \delta^\alpha_\beta + \Gamma^{\alpha}_{\mu\kappa}\delta^\kappa_\beta - \Gamma^{\kappa}_{\mu\beta}\delta^\alpha_\kappa =$

$$= 0 + \Gamma^{\alpha}_{\mu\beta} - \Gamma^{\alpha}_{\mu\beta} = 0 \qquad 4.65$$

  as might be expected.

- $\tilde{\nabla}\mathbf{g} \overset{comp}{\rightarrow} \nabla_\mu g_{\alpha\beta} = \partial_\mu g_{\alpha\beta} - \Gamma^{\kappa}_{\mu\alpha} g_{\kappa\beta} - \Gamma^{\kappa}_{\mu\beta} g_{\alpha\kappa} = 0 \qquad 4.66$

  ▫ Indeed (*eq.2.37, eq.4.30*):

$$\partial_\mu g_{\alpha\beta} = \partial_\mu(\vec{e}_\alpha\cdot\vec{e}_\beta) = \underbrace{\partial_\mu\vec{e}_\alpha}_{=\Gamma^{\kappa}_{\mu\alpha}\vec{e}_\kappa}\cdot\vec{e}_\beta + \vec{e}_\alpha\cdot\underbrace{\partial_\mu\vec{e}_\beta}_{=\Gamma^{\kappa}_{\mu\beta}\vec{e}_\kappa} =$$

$$= \Gamma^{\kappa}_{\mu\alpha}\vec{e}_\kappa\cdot\vec{e}_\beta + \vec{e}_\alpha\cdot\Gamma^{\kappa}_{\mu\beta}\vec{e}_\kappa = \Gamma^{\kappa}_{\mu\alpha} g_{\kappa\beta} + \Gamma^{\kappa}_{\mu\beta} g_{\alpha\kappa} \quad *$$

This important result, less obvious than the previous one, states that

| in each point of any manifold with Riemannian metric it is:

$$\tilde{\nabla}\mathbf{g} = 0 \qquad 4.67$$

Likewise:

$$\tilde{\nabla}\mathbf{g}^{-1} = 0 \qquad 4.68$$

---

* $\tilde{\nabla}\mathbf{g} = 0$ in every point does *not* mean that $\mathbf{g}$ is constant, but *stationary* in covariant sense, variable at least because of the second derivatives.

▫ It can be obtained like *eq.4.66*, passing through $\nabla_\mu g^{\alpha\beta}$ and $\partial_\mu g^{\alpha\beta}$ and then considering that $\partial_\mu \tilde{e}^\alpha = -\Gamma^\alpha_{\mu\kappa} \tilde{e}^\kappa$ (*eq.4.39*).

The last *eq.4.67*, *eq.4.68* are often respectively written as:

$$\nabla_\kappa g_{\alpha\beta} = 0 \quad \text{or} \quad g_{\alpha\beta;\kappa} = 0$$

and:

$$\nabla_\kappa g^{\mu\nu} = 0 \quad \text{or} \quad g^{\mu\nu}{}_{;\kappa} = 0 \qquad 4.69$$

## 4.17 Covariant derivative and index raising / lowering

The covariant derivative and the raising or lowering of the indexes are operations that commute (= you can reverse the order in which they take place). For example:

$$g^{\alpha\mu} T_{\alpha\beta;\kappa} = T^\mu_{\beta;\kappa} \qquad 4.70$$

(on the left, covariant derivative followed by raising the index; on the right, raised index followed by covariant derivative).

▫ Indeed:

$$T^\mu_{\beta;\kappa} = \left(g^{\alpha\mu} T_{\alpha\beta}\right)_{;\kappa} = \underbrace{g^{\alpha\mu}{}_{;\kappa}}_{=0} T_{\alpha\beta} + g^{\alpha\mu} T_{\alpha\beta;\kappa} = g^{\alpha\mu} T_{\alpha\beta;\kappa}$$

having used the *eq.4.69*.

It is thus allowed "to raise/lower indexes under covariant derivative".

## 4.18 Christoffel symbols

Christoffel symbols $\Gamma^\lambda_{\mu\nu}$ give account of the variability of basis-vectors from one point to the other of the manifold: as already noted, if basis-vectors are constant, all $\Gamma$ are null.

▪ In the usual 3D Euclidean space described by a Cartesian coordinate system, basis-vectors do not vary from point to point and consequently $\forall\,\Gamma = 0$ everywhere (so that covariant derivatives coincide with partial derivatives).

That is no more true when the same space is described by a spherical coordinate system $\rho, \phi, \theta$ because in this case the basis-vectors $\vec{e}_\rho, \vec{e}_\phi, \vec{e}_\theta$ are functions of the point (as we saw for the plane polar coordinates) and this implies that at least some $\Gamma$ are $\neq 0$.

Just this fact makes sure that the $\Gamma^\lambda_{\mu\nu}$ cannot be the components of a hypothetical tensor $\binom{1}{2}$ as might seem: the hypothetical tensor $\mathbf{\Gamma}$

would be null in Cartesian (since all its components are zero) and not null in spherical coordinates, against the invariance of a tensor under change of coordinates.
The Christoffel symbols are rather a set of $n^3$ coefficients depending on 3 indexes but do not form any tensor. However, observing the way in which it was introduced $\left(\vec{\Gamma} = \Gamma^{\lambda} \vec{e}_{\lambda} \ , \ eq.4.29\right)$ , each $\vec{\Gamma}$ related to a fixed pair of lower indexes is a vector whose components are marked by the upper index $\lambda$. It is therefore valid (only for this index) raising / lowering by means of **g**:

$$g_{\kappa\lambda} \Gamma^{\lambda}_{\mu\nu} = \Gamma_{\mu\nu\kappa} \qquad 4.71$$

▪ An important property of the Christoffel symbols is to be symmetric *in a coordinate basis* with respect to the exchange of lower indexes:

$$\Gamma^{\lambda}_{\mu\nu} = \Gamma^{\lambda}_{\nu\mu} \qquad 4.72$$

▫ To prove this we operate a change of coordinates and the consequent transformation of coordinate basis by the appropriate related matrix. Let us express the old $\Gamma^{\lambda}_{\mu\nu}$ in terms of the new basis (using for this purpose the matrix $\Lambda^{\kappa'}_{\nu} = \frac{\partial x^{\kappa'}}{\partial x^{\nu}}$) :

$$\Gamma^{\lambda}_{\mu\nu} \vec{e}_{\lambda} \stackrel{\text{def}}{=} \partial_{\mu} \vec{e}_{\nu} = \partial_{\mu}\left(\Lambda^{\kappa'}_{\nu} \vec{e}_{\kappa'}\right) = \left(\partial_{\mu} \Lambda^{\kappa'}_{\nu}\right) \vec{e}_{\kappa'} + \Lambda^{\kappa'}_{\nu}\left(\partial_{\mu} \vec{e}_{\kappa'}\right) =$$

$$= \partial_{\mu} \frac{\partial x^{\kappa'}}{\partial x^{\nu}} \vec{e}_{\kappa'} + \frac{\partial x^{\kappa'}}{\partial x^{\nu}} \partial_{\mu} \vec{e}_{\kappa'} = \frac{\partial^2 x^{\kappa'}}{\partial x^{\mu} \partial x^{\nu}} \vec{e}_{\kappa'} + \frac{\partial x^{\kappa'}}{\partial x^{\nu}} \frac{\partial x^{\lambda'}}{\partial x^{\mu}} \frac{\partial}{\partial x^{\lambda'}} \vec{e}_{\kappa'}$$

where we have used the chain rule to insert $x^{\lambda'}$.

Since this expression is symmetric with respect to the indexes $\mu$, $\nu$, the same result is obtained by calculating $\Gamma^{\lambda}_{\nu\mu} \vec{e}_{\lambda} \stackrel{\text{def}}{=} \partial_{\nu} \vec{e}_{\mu}$ on the new basis, then $\Gamma^{\lambda}_{\mu\nu} = \Gamma^{\lambda}_{\nu\mu}$ .

▪ To get a better understanding of the meaning of connection coefficients $\Gamma$ and related indexes, let's write *eq.4.30* $\partial_{\mu} \vec{e}_{\nu} = \Gamma^{\lambda}_{\mu\nu} \vec{e}_{\lambda}$ in full:

$$\partial_1 \vec{e}_1 = \Gamma^1_{11} \vec{e}_1 + \Gamma^2_{11} \vec{e}_2 + \Gamma^3_{11} \vec{e}_3 + \ldots$$

$$\partial_1 \vec{e}_2 = \Gamma^1_{12} \vec{e}_1 + \Gamma^2_{12} \vec{e}_2 + \Gamma^3_{12} \vec{e}_3 + \ldots$$

. . . . . . . . .

$$\partial_2 \vec{e}_1 = \Gamma^1_{21} \vec{e}_1 + \Gamma^2_{21} \vec{e}_2 + \underline{\Gamma^3_{21}} \vec{e}_3 + \ldots$$

and take in consideration, for instance, the coefficient $\Gamma^{3}_{21}$ from the last equation. By observing the role it plays in the equation where it appears, it can be understood as:

$$\Gamma^{3}_{21}$$

3 ← component along $\vec{e}_3$ ...
↓ ...of the derivative respect to $x^2$...
2 1 ← ...of the basis-vector $\vec{e}_1$

In general:

> $\Gamma^{\lambda}_{\mu\nu}$ can be understood as "component along $\vec{e}_\lambda$ of the derivative $\partial_\mu$ of the basis-vector $\vec{e}_\nu$ ".

The whole story is: moving a small step $d\vec{x}$ from an initial point, all basis-vectors change; among others the ν-th basis-vector $\vec{e}_\nu$ will change; $n$ partial derivatives will describe its variation along each of the $n$ coordinate lines; in particular $\partial_\mu \vec{e}_\nu$ will stand for its variation rate along the coordinate line $x^\mu$; $\partial_\mu \vec{e}_\nu$ is a vector that can be decomposed into $n$ components in direction of each basis-vectors. We denote $\Gamma^{\lambda}_{\mu\nu}$ the component in the direction of $\vec{e}_\lambda$.

▪ An important relationship links Γ to **g** and shows that *in coordinate bases* the coefficients Γ are functions of **g** and its first derivatives only :

$$\Gamma^{\kappa}_{\mu\beta} = \frac{1}{2} g^{\kappa\alpha}(-g_{\mu\beta,\alpha} + g_{\beta\alpha,\mu} + g_{\alpha\mu,\beta}) \qquad 4.73$$

(note the cyclicity of indexes μ, β, α within the parenthesis).

▫ It follows from the Riemann spaces property *eq.4.67*::

$$\tilde{\nabla}\mathbf{g}=0 \;\Rightarrow\; \nabla_\mu g_{\alpha\beta} = \partial_\mu g_{\alpha\beta} - \Gamma^{\lambda}_{\mu\alpha} g_{\lambda\beta} - \Gamma^{\lambda}_{\mu\beta} g_{\alpha\lambda} = 0$$

⇒ (using the comma instead of ∂ as derivative symbol) :

$$g_{\alpha\beta,\mu} = \Gamma^{\lambda}_{\mu\alpha} g_{\lambda\beta} + \Gamma^{\lambda}_{\mu\beta} g_{\alpha\lambda} \qquad 4.74$$

Cyclically rotating the three indexes α, β, μ we can write two other similar equations for $g_{\beta\mu,\alpha} = ....$ and $g_{\mu\alpha,\beta} = ....$

By summing member to member: 1st – 2nd + 3rd equation and using the symmetry properties of the lower indexes *eq.4.72* (which applies in a coordinate basis) we get:

$$g_{\alpha\beta,\mu} - g_{\beta\mu,\alpha} + g_{\mu\alpha,\beta} = 2\, g_{\lambda\alpha} \Gamma^{\lambda}_{\mu\beta}$$

Multiplying both members by $g^{\kappa\alpha}$ and since $g^{\kappa\alpha} g_{\lambda\alpha} = \delta^{\kappa}_{\lambda}$ : *

$$\Gamma^{\kappa}_{\mu\beta} = \frac{1}{2} g^{\kappa\alpha} \left( g_{\alpha\beta,\mu} - g_{\beta\mu,\alpha} + g_{\mu\alpha,\beta} \right) \quad \text{, c.v.}$$

This result has as important consequence:**

$$\forall\, g_{\alpha\beta} = constant \quad \Leftrightarrow \quad \forall\, \Gamma = 0 \qquad 4.75$$

> If the components $g_{\alpha\beta}$ of the metric tensor **g** are constant, all the coefficients $\Gamma$ are zero, and vice versa.

This is especially true in Cartesian coordinate systems.

▪ The following properties constitute an indivisible "take or leave" block:

$$\begin{cases} \forall\, \vec{e}_{\mu} = costant \\ \forall\, g_{\mu\nu} = costant \ \ \text{(i.e. } \mathbf{g} = \text{costant)} \\ \forall\, \Gamma = 0 \\ \nabla_{\mu} = \partial_{\mu} \ \ \text{(covariant derivatives = partial derivatives)} \end{cases}$$

in the sense that, for a given space, all are valid or none.

▫ In fact all propositions make a single indivisible block because they imply each other, as can be easily seen by *eq.2.39*, *eq.4.31*, *eq.4.47*, *eq.4.73*.

In general we will qualify "flat" a metric where all these properties hold (particularly $\forall\, \Gamma = 0$). ***

---

* From *eq.2.28*, since **g** and **g**$^{-1}$ are inverse tensors.

** Which is already implicit in *eq.2.39* and *eq.4.31*.

***Even if flatness (or not) should be a feature of the space. Note that flat metric $\Rightarrow$ flat space, but a space can be flat without the metric being flat. In any case, flat space $\Rightarrow \exists$ a coordinate system in which the properties in question are verified, as we will further on.

## 4.19 Covariant derivative and invariance of tensor equations

A tensor equation contains only tensors, thus it may contain covariant derivatives but not ordinary derivatives (which are not tensors). For instance, $\tilde{\nabla} \mathbf{g} = 0$ or its componentwise equivalent $\nabla_\mu g_{\alpha\beta} = 0$ (which actually stands for $n^3$ equations) are tensor equations; it is *not* the case of *eq.4.73*.

A strategy often used to obtain tensor equations is called "*comma goes to semi-colon*" and takes advantage of the fact that in a "flat" coordinate system, the ordinary derivative and the covariant derivative coincide since all Γ are zero. In practice:

> Working in a "flat" coordinate system (e.g., Cartesian), a partial derivatives equation that proves to be valid in this system can be turned into a tensor equation simply by replacing partial derivatives with covariant derivatives, i.e. by replacing commas with semicolons. The tensor equation obtained in this way is invariant and applies in any reference accessible via coordinate transformation.*

The usefulness of such a strategy is also due to the fact that any curved space is locally flat in each point (provided it is not "pathological"), as will be seen further on.

* To avoid possible misunderstandings it is worth to clarify that saying "in any reference" does not mean "in any space": the validity of the tensorial equation is limited to the particular space you deal with and does not extend outside to other spaces.

## 4.20 T-mosaic representation of gradient, divergence and covariant derivative

$\tilde{\nabla}$ is the covector $\boxed{\nabla_\mu \atop \tilde{e}^\mu}$ that applies to scalars, vectors or tensors.

- Gradient of scalar: $\quad \tilde{\nabla} f \;=\; \boxed{\nabla_\mu \; f \atop \tilde{e}^\mu} \;=\; \boxed{\nabla_\mu f \atop \tilde{e}^\mu}$

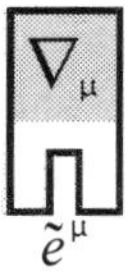

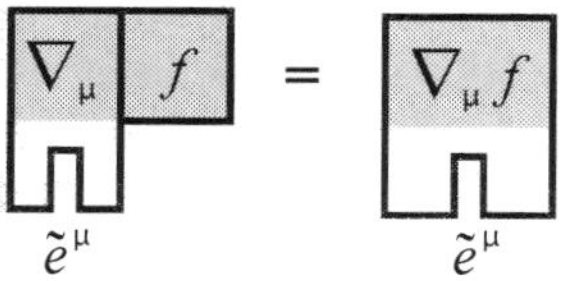

- Gradient of vector: $\quad \tilde{\nabla} \otimes \vec{V} \;=\; \boxed{\nabla_\mu \; V^\nu \atop \tilde{e}^\mu}^{\vec{e}_\nu} \;=\; \boxed{\nabla_\mu V^\nu \atop \tilde{e}^\mu}^{\vec{e}_\nu} \;=$

$$= \underbrace{\nabla_\mu V^\nu}_{\text{covariant derivative}} \; \tilde{e}^\mu \vec{e}_\nu$$

- Gradient of covector: $\quad \tilde{\nabla} \otimes \tilde{P} \;=\; \boxed{\nabla_\mu \; P_\nu \atop \tilde{e}^\mu \; \tilde{e}^\nu} \;=\; \boxed{\nabla_\mu P_\nu \atop \tilde{e}^\mu \; \tilde{e}^\nu} \;=$

$$= \underbrace{\nabla_\mu P_\nu}_{\text{covariant derivative}} \; \tilde{e}^\mu \tilde{e}^\nu$$

- Gradient of tensor,

  e.g. **g**: $\quad \tilde{\nabla} \otimes \mathbf{g} \;=\; \boxed{\nabla_\lambda \; g_{\mu\nu} \atop \tilde{e}^\lambda \; \tilde{e}^\mu \; \tilde{e}^\nu} \;=\; \boxed{\nabla_\lambda g_{\mu\nu} \atop \tilde{e}^\lambda \; \tilde{e}^\mu \; \tilde{e}^\nu}$

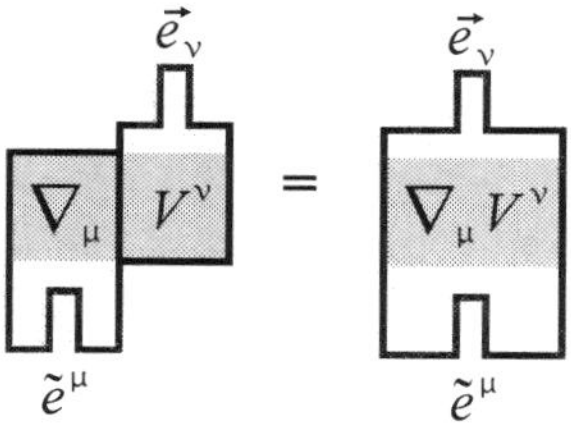

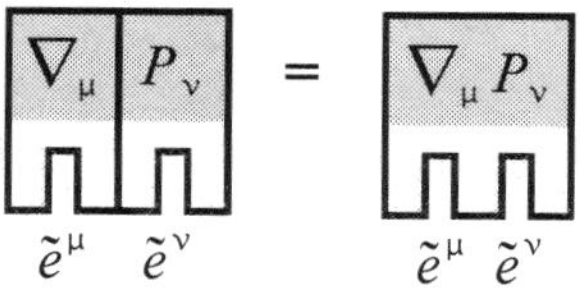

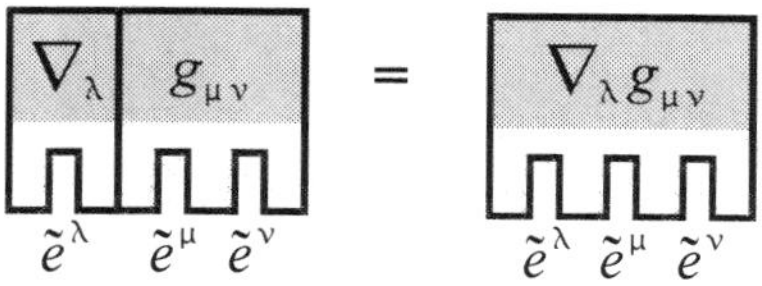

- Divergence of vector:

$$div\vec{A} = \tilde{\nabla}(\vec{A}) = \quad \rightarrow \quad = \nabla_\mu A^\mu$$

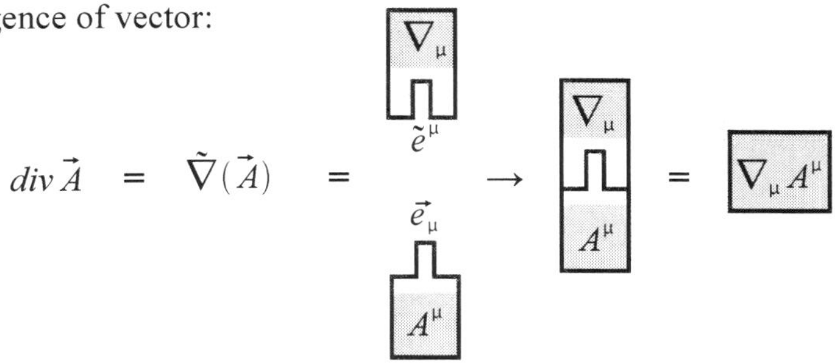

- Divergence of tensor $\mathbf{T} = T_\lambda^{\nu\mu}\vec{e}_\nu\vec{e}_\mu\tilde{e}^\lambda$ (with respect to index $\mu$):

$$\tilde{\nabla}(\mathbf{T}) = \quad \rightarrow \quad = \nabla_\mu T_\lambda^{\nu\mu}$$

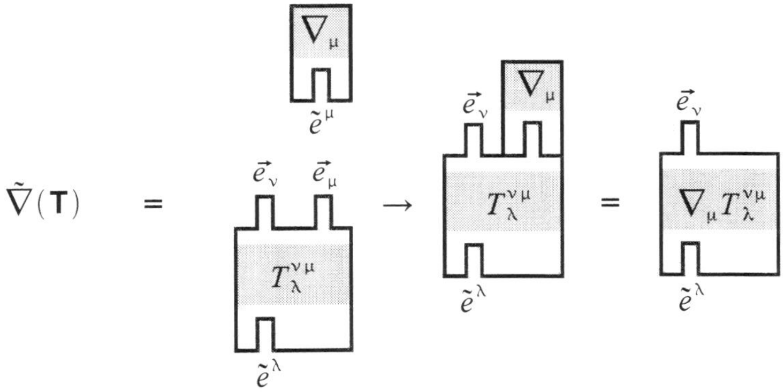

# 5 Curved manifolds

## 5.1 Symptoms of curvature

Consider a spherical surface within the usual 3D Euclidean space. The spherical surface is a manifold 2D. A 3D observer, or $\mathcal{O}_{3D}$, which has an outside view, sees the curvature of the spherical surface in the third dimension. A 2D observer $\mathcal{O}_{2D}$ who lives on the spherical surface has no outside view. $\mathcal{O}_{3D}$ sees light rays propagate on the spherical surface along maximum circle arcs; for $\mathcal{O}_{2D}$ they will be straight lines by definition (they are the only possible inertial trajectories, run in absence of forces). How can $\mathcal{O}_{2D}$ realize that his space is curved?
By studying geometry in its local context $\mathcal{O}_{2D}$ will discover the usual properties of the 2D Euclidean geometry, or plane geometry. For example, he will finds that the ratio of the circumference to diameter is a constant = $\pi$ for all circles and that the sum of the inner angles of any triangle is 180°. In fact, in its environment, i.e. in the small portion of the spherical surface on which his life takes place, his two-dimensional space is locally flat, as for us is flat the surface of a water pool on the Earth's surface (the observer $\mathcal{O}_{3D}$ sees this neighborhood practically lying on the plane tangent to the spherical surface at the point where $\mathcal{O}_{2D}$ is located).
However, when $\mathcal{O}_{2D}$ comes to consider very large circles, he realizes that the circumference / diameter ratio is no more a constant and it is smaller and smaller than $\pi$ as the circle enlarges, and that the sum of the inner angles of a triangle is variable but always > 180°.
$\mathcal{O}_{2D}$ can deduce from these facts that his space is curved, though not capable to understand the third dimension.
The curvature of a space is thus an intrinsic property of the space itself and there is no need of an outside view to describe it. In other words, it is not necessary for a $n$-dimensional space to be considered embedded in another $n+1$ dimensional space to reveal the curvature.

▪ Another circumstance that allows $\mathcal{O}_{2D}$ to discover the curvature of his space is that, carrying a vector parallel to itself on a large enough closed loop, the returned vector is no longer parallel ( = does not overlap) to the initial vector.
In a flat space such as the Euclidean a vectors can be transported parallel to itself without difficulty, so that they can be often considered

delocalized. For example, to measure the relative velocity of two particles far apart, we may imagine to transport the vector $\vec{v}_2$ parallel to itself from the second particle to the first so as to make their origins coincide in order to carry out the subtraction $\vec{v}_2 - \vec{v}_1$ .
But what does it mean a parallel transport in a curved space?
In a curved space, this expression has no precise meaning. However, $\mathcal{O}_{2D}$ can think to perform a parallel transport of a vector by a step-by-step strategy, taking advantage from the fact that in the neighborhood of each point his space looks substantially flat.

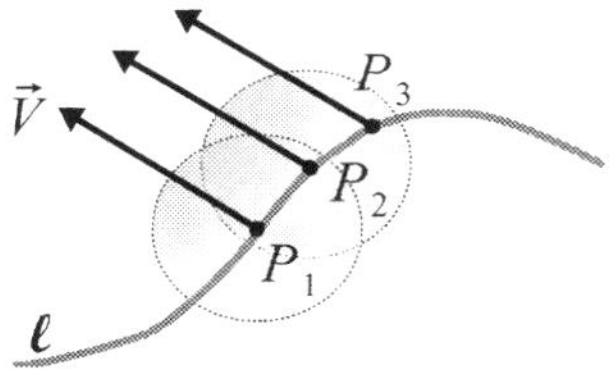


Let $\ell$ be the curve along which he wants to parallel-transport the (foot of the) vector $\vec{V}$ . At first the observer $\mathcal{O}_{2D}$ and the vector are in $P_1$ ; the vector is transported parallel to itself to $P_2$ , a point of the (infinitesimal) neighborhood of $P_1$ . Then $\mathcal{O}_{2D}$ goes to $P_2$ and parallel transports the vector to $P_3$ that belongs to the neighborhood of $P_2$ , and so on. In this way the transport is always within a neighborhood in which the space looks flat and the notion of parallel transport is not ambiguous: vector in $P_1$ // vector in $P_2$ ; vector in $P_2$ // vector in $P_3$ , and so on.
But the question is: is vector in $P_1$ // vector in $P_n$ ? That is: does the parallelism which works locally step-by-step by infinitesimal amounts hold as well globally?
To answer this question we ought to transport the vector along a closed path: only if the vector that returns from parallel transport is superimposable onto the initial vector we can conclude for global parallelism to have been preserved.
In fact it is not so, at least for large enough circuits: just take for example the transport *A*-*B*-*C*-*A* along the meridians and the equator on

a spherical surface:

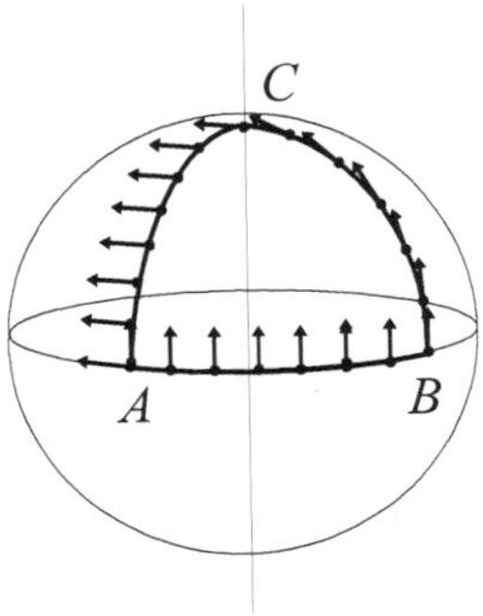


The mismatch between the initial vector and the one that returns after the parallel transport can give a measure of the degree of curvature of the space (this measure is fully accessible to the two-dimensional observer $\mathcal{O}_{2D}$).
To do that quantitatively we'll give a more precise mathematical definition of parallel transport of a vector along a line further on.

▪ A more detailed explanation of how parallel transport of a vector works in a curved space can be given in terms of tangent plane; for that it is necessary to think the space in question as embedded in a space with an extra-dimension. We'll refer here to a transport on a 2D curved surface, as seen by a $\mathcal{O}_{3D}$ observer.

Let a vector $\vec{V}$ be defined at a point $P$ of the surface, to be transported along a line $\ell$ defined on the surface.
The transport of the vector $\vec{V}$ on the curved surface along the line $\ell$ starting from the point $P$ works as follows:
• Let us take the tangent plane to the surface at $P$ .
• If you move the plane keeping it in contact with the curved surface without slipping, the tangency point will move too. So, we'll move the plane so that the point of tangency goes forward *continously* along the curve $\ell$.
• During this route, from each instantaneous point of tangency, we draw *on the plan* a vector parallel to the initial one $\vec{V}$. Since the plane and the curved surface coincide in the tangency point, the vector belongs *to both them*. We imagine to draw it on the surface too.
• At the end of the journey the plane will keep track of the line $\ell$ (it is a

different curve because flattened). In each point of this line we'll find drawn a vector pointing outward; all these vectors are parallel to each other and to the initial vector $\vec{V}$.

• On the curved surface too we will find a vector coming out from each point of the line $\ell$, all them equal to $\vec{V}$ in magnitude but *not in direction*: all vectors drawn on the curved surface will have a different orientation to one another and different from $\vec{V}$. In particular, for a closed path on the surface: $\vec{V}_{fin} \neq \vec{V}_{init}$ .

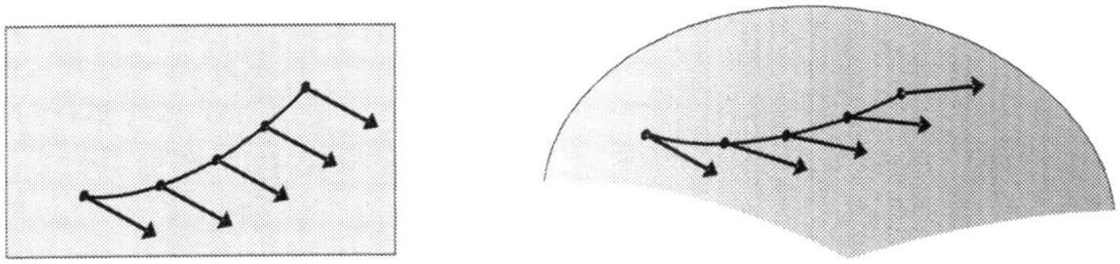

The reason is that, along the way, the tangent plane has been continuously changing its spatial orientation to adhere to the surface all along the line, describing complex rotations, so that the vector, always parallel to itself in the plane, progressively changes its direction on the surface.

This modality of moving the vector $\vec{V}$ on the curved surface is what is meant by "parallel transport" of a vector along a line on a curved surface (although the parallelism is only on the tangent mobile plan, not on the curved surface).

It is worth noting that, in the neighborhood of the tangency point, the distance between the curved surface and the tangent plane is an infinitesimal of higher order than the displacement from the tangency point; hence also the plan orientation change and the variation of the vector are infinitesimal of higher order than the displacement along $\ell$.

We also note that cutting out of the curved surface a *narrow strip* containing the line, extending it on a plane, and associating to each point a vector parallel to $\vec{V}_{init}$ , we get a figure that is identical and superimposable to that one drawn on the mobile plan. The two procedures are therefore equivalent. We'll use the latter further on.

All that applies to a $\mathcal{O}_{3D}$ observer; a $\mathcal{O}_{2D}$ observer has no notion of a tangent plane and, as mentioned before, will carry the vector parallel to itself without any problems, point after point, in his locally flat 2D space, as if acting in a plane (except finding a different-oriented vector at the end of a closed circuit).

## 5.2 Derivative of a scalar or vector along a line

Given a scalar or a vector defined (at least at the points of a parametric line) within the manifold, let us express the derivatives of the scalar or vector along the line as the incremental ratio of the scalar or vector with respect to the parameter.

The line $\ell$ is defined by $n$ parametric equations $x^{\mu}(\tau)$. The parameter $\tau$ can be (or not) the arc length. At each point of the line let be defined a tangent vector (a unit vector only if $\tau$ is the arc):

$$\vec{U} = \frac{d\vec{x}}{d\tau} \qquad 5.1$$

$$\Rightarrow \quad U^{\mu} = \frac{dx^{\mu}}{d\tau} \quad \text{i.e.} \quad \vec{U} \overset{comp}{\rightarrow} \frac{dx^{\mu}}{d\tau} \quad \textit{in coordinate basis.}$$

### ❶ derivative of a scalar $f$ along a line

From the total differential of $f$: $df = \frac{\partial f}{\partial x^{\mu}} dx^{\mu}$ we immediately deduce the definition of derivative along a line:

$$\frac{df}{d\tau} = \frac{\partial f}{\partial x^{\mu}} \frac{dx^{\mu}}{d\tau} \qquad 5.2$$

and, since for a scalar $f$ partial and covariant derivatives identify, *in a coordinate base* is:

$$\frac{df}{d\tau} = U^{\mu}\partial_{\mu}f \equiv U^{\mu}\nabla_{\mu}f = \langle \tilde{\nabla} f, \vec{U} \rangle \qquad 5.3$$

Roughly speaking: the derivative of $f$ along the line can be seen as a "*projection*" or a "*component*" *of the gradient in direction of the tangent vector.* It is a scalar.

▪ The derivative of a scalar along a line is a generalization of the partial $\partial_{\mu}$ (or covariant $\nabla_{\mu}$ ) derivative and reduces to it along coordinated lines.

- Indeed: along a coordinate line $x^{\bar{\mu}}$ it is $\vec{U} \,//\, \vec{e}_{\bar{\mu}} \Rightarrow U^{\bar{\mu}}$ is the only nonzero component; * the summation on $\mu$ implicit in $U^{\mu}\,\nabla_{\mu} f$ (*eq.5.3*) thus collapses to the single $\bar{\mu}$-th term.

  Moreover, if $\tau$ is the arc, then $d\tau \equiv dx^{\bar{\mu}} \Rightarrow U^{\bar{\mu}} = \dfrac{dx^{\bar{\mu}}}{d\tau} = 1$ .

  For these reasons, *eq.5.3* reduces to:

$$\frac{df}{d\tau} = U^{\bar{\mu}} \nabla_{\bar{\mu}} f \,[\text{no sum}] = \nabla_{\bar{\mu}} f \quad ** \tag{5.4}$$

q.e.d.

▪ As a projection on $\vec{U}$ of the gradient $\nabla f$ the derivative of a scalar along a line is also denoted:

$$\frac{df}{d\tau} \equiv \nabla_{\vec{U}} f \tag{5.5}$$

▪ *Eq.5.3* and *eq.5.5* enable us to write symbolically (for a scalar):

$$\frac{d}{d\tau} = \nabla_{\vec{U}} = U^{\mu} \partial_{\mu} = U^{\mu} \nabla_{\mu} \tag{5.6}$$

## ❷ derivative of a vector $\vec{V}$ along a line

In analogy with the result just found for the scalar $f$ (*eq.5.3*) we define the derivative of a vector $\vec{V}$ along a line $\ell$ whose tangent vector is $\vec{U}$ as the *"projection" or "component" of the gradient on the tangent vector*:

$$\frac{d\vec{V}}{d\tau} \stackrel{\text{def}}{=} \langle \tilde{\nabla} \vec{V}, \vec{U} \rangle \tag{5.7}$$

The derivative of a vector along a line is a vector, since $\tilde{\nabla}\vec{V}$ is a $\binom{1}{1}$ tensor. Expanding on the bases:

$$\frac{d\vec{V}}{d\tau} = \langle \tilde{\nabla}\vec{V}, \vec{U} \rangle = \langle \nabla_{\mu} V^{\nu} \tilde{e}^{\mu} \vec{e}_{\nu}, U^{\kappa} \vec{e}_{\kappa} \rangle = \nabla_{\mu} V^{\nu}\, U^{\kappa} \underbrace{\langle \tilde{e}^{\mu}, \vec{e}_{\kappa} \rangle}_{\delta^{\mu}_{\kappa}} \vec{e}_{\nu} =$$

---

* Note that along the coordinate line $x^{\bar{\mu}}$, provided $\tau$ is the arc:

$$\vec{U} \stackrel{\text{def}}{=} \frac{d\vec{x}}{d\tau} = \frac{\vec{e}_{\mu}\, dx^{\mu}}{d\tau} = \frac{\vec{e}_{\bar{\mu}}\, dx^{\bar{\mu}}}{dx^{\bar{\mu}}} [\text{no sum}] = \vec{e}_{\bar{\mu}}$$

** The indication [no sum] offs the summation convention for repeated indexes in the case.

$$= \underbrace{U^{\mu} \nabla_{\mu} V^{\nu}}_{\overset{\text{def}}{\equiv} \frac{DV^{\nu}}{d\tau}} \vec{e}_{\nu} = \frac{DV^{\nu}}{d\tau} \vec{e}_{\nu} \qquad 5.8$$

i.e.
$$\frac{d\vec{V}}{d\tau} \overset{comp}{\rightarrow} \frac{DV^{\nu}}{d\tau} \qquad 5.9$$

after set:
$$\frac{DV^{\nu}}{d\tau} \overset{\text{def}}{\equiv} U^{\mu} \nabla_{\mu} V^{\nu} \qquad 5.10$$

$\frac{DV^{\nu}}{d\tau}$ is the $\nu$-component of the derivative of vector $\frac{d\vec{V}}{d\tau}$ and is referred to as *covariant derivative along line* or *directional covariant derivative* or *absolute derivative*; it can be expressed as:

$$\frac{DV^{\nu}}{d\tau} \overset{\text{def}}{\equiv} U^{\mu} \nabla_{\mu} V^{\nu} = \underbrace{U^{\mu}}_{= \frac{dx^{\mu}}{d\tau}} \left( \frac{\partial V^{\nu}}{\partial x^{\mu}} + \Gamma^{\nu}_{\mu\lambda} V^{\lambda} \right) = \frac{\partial V^{\nu}}{\partial x^{\mu}} \frac{dx^{\mu}}{d\tau} + \Gamma^{\nu}_{\mu\lambda} V^{\lambda} U^{\mu} =$$
$$= \frac{dV^{\nu}}{d\tau} + \Gamma^{\nu}_{\mu\lambda} V^{\lambda} U^{\mu} \qquad 5.11$$

A new derivative symbol has been introduced since the components of $\frac{d\vec{V}}{d\tau}$ are *not* simply $\frac{dV^{\nu}}{d\tau}$, but take the more complex form of *eq.5.11* (only when $\forall \Gamma = 0$ expressions $\frac{DV^{\nu}}{d\tau}$ and $\frac{dV^{\nu}}{d\tau}$ are equivalent).

▪ As a projection on $\vec{U}$ of the gradient $\nabla \vec{V}$ the derivative of a vector along a line is also denoted:

$$\frac{d\vec{V}}{d\tau} \equiv \nabla_{\vec{U}} \vec{V} \qquad 5.12$$

▪ Let us recapitulate the various notations for the derivative of a vector along a line and for its components. Beside the definition *eq.5.7* $\frac{d\vec{V}}{d\tau} \overset{\text{def}}{\equiv} \langle \tilde{\nabla} \vec{V}, \vec{U} \rangle$ , *eq.5.12*, *eq.5.9*, *eq.5.10* have been appended, summarized in:

$$\frac{d\vec{V}}{d\tau} \equiv \nabla_{\vec{U}} \vec{V} \overset{comp}{\rightarrow} \frac{DV^{\nu}}{d\tau} \equiv U^{\mu} \nabla_{\mu} V^{\nu} \qquad 5.13$$

or, symbolically:

$$\frac{d}{d\tau} \equiv \nabla_{\vec{U}} \quad \overset{comp}{\rightarrow} \quad \frac{D}{d\tau} \equiv U^{\mu}\nabla_{\mu} \qquad 5.14$$

▪ The relationship between the derivative of a vector along a line and the covariant derivative becomes apparent when the line $\ell$ is a coordinate line $x^{\bar{\mu}}$.

▫ In this case $U^{\bar{\mu}} = 1$ provided $\tau$ is the arc and the summation on $\bar{\mu}$ collapses, as already seen (*eq.5.4*) for the derivative of a scalar; then the chain *eq.5.8* changes into:

$$\frac{d\vec{V}}{d\tau} = U^{\bar{\mu}}\nabla_{\bar{\mu}}V^{\nu}\vec{e}_{\nu} = \nabla_{\bar{\mu}}V^{\nu}\vec{e}_{\nu} \qquad 5.15$$

namely: $$\frac{d\vec{V}}{d\tau} \overset{comp}{\rightarrow} \nabla_{\bar{\mu}}V^{\nu}$$

The derivative of a vector along a coordinate line $x^{\bar{\mu}}$ has as components the $n$ covariant derivatives corresponding to the blocked index $\bar{\mu}$.

*Notation*

Derivatives along a line of :

• scalar $f$: $$\frac{df}{d\tau} = \langle \tilde{\nabla} f, \vec{U} \rangle = \nabla_{\vec{U}} f = U^{\mu}\nabla_{\mu} f \quad (= U^{\mu}\partial_{\mu} f)$$

• vector $\vec{V}$: $$\frac{d\vec{V}}{d\tau} = \langle \tilde{\nabla}\vec{V}, \vec{U} \rangle = \nabla_{\vec{U}}\vec{V}$$

$$\overset{comp}{\longrightarrow} \frac{DV^{\mu}}{d\tau} = U^{\mu}\nabla_{\mu}V^{\nu} \quad (= U^{\mu}V^{\nu}_{\ ;\mu})$$

▪ Comparing the two cases of derivative along a line of a scalar and of a vector we see (*eq.5.6*, *eq.5.14*) that in both cases the derivative is denoted by $\frac{d}{d\tau} = \nabla_{\vec{U}}$ but the operator $\nabla_{\vec{U}}$ has a different meaning:

- $\nabla_{\vec{U}} = U^{\mu}\nabla_{\mu}$ for a scalar
- $\nabla_{\vec{U}} \overset{comp}{\rightarrow} U^{\mu}\nabla_{\mu}$ for a vector.

## 5.3 T-mosaic representation of derivatives along a line

❶ Derivative along a line of a scalar:

$$\frac{df}{d\tau} = \langle \tilde{\nabla} f, \vec{U} \rangle = \boxed{\nabla_\mu f \;/\; \tilde{e}^\mu \;/\; \vec{e}_\mu \;/\; U^\mu} \rightarrow \boxed{\nabla_\mu f \;/\; U^\mu} = \boxed{U^\mu \nabla_\mu f} =$$

$$= \underbrace{\boxed{\nabla_{\tilde{U}} f}}_{\text{(scalar)}}$$

❷ Derivative along a line of a vector:

$$\frac{d\vec{V}}{d\tau} = \langle \tilde{\nabla} \vec{V}, \vec{U} \rangle = \boxed{\nabla_\mu V^\nu \;/\; \vec{e}_\nu \;/\; \tilde{e}^\mu \;/\; \vec{e}_\mu \;/\; U^\mu} \rightarrow \boxed{\nabla_\mu V^\nu \;/\; \vec{e}_\nu \;/\; \tilde{e}^\mu \;/\; \vec{e}_\mu \;/\; U^\mu} = \boxed{U^\mu \nabla_\mu V^\nu \;/\; \vec{e}_\nu} =$$

$$= \boxed{\frac{DV^\nu}{d\tau} \;/\; \vec{e}_\nu} \equiv \underbrace{\nabla_{\tilde{U}} \vec{V}}_{\text{(vector)}}$$

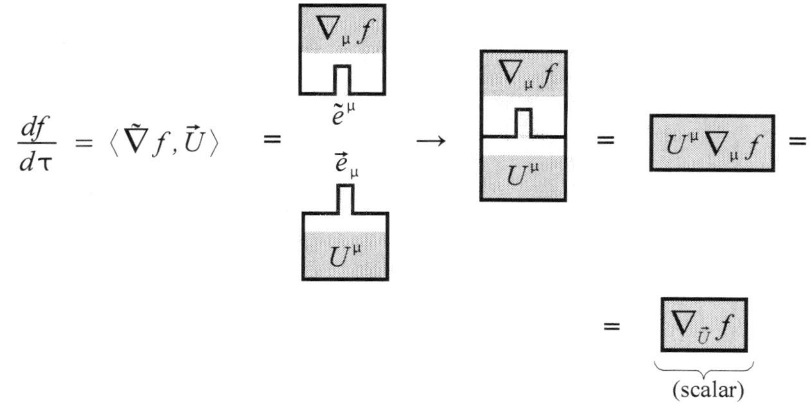

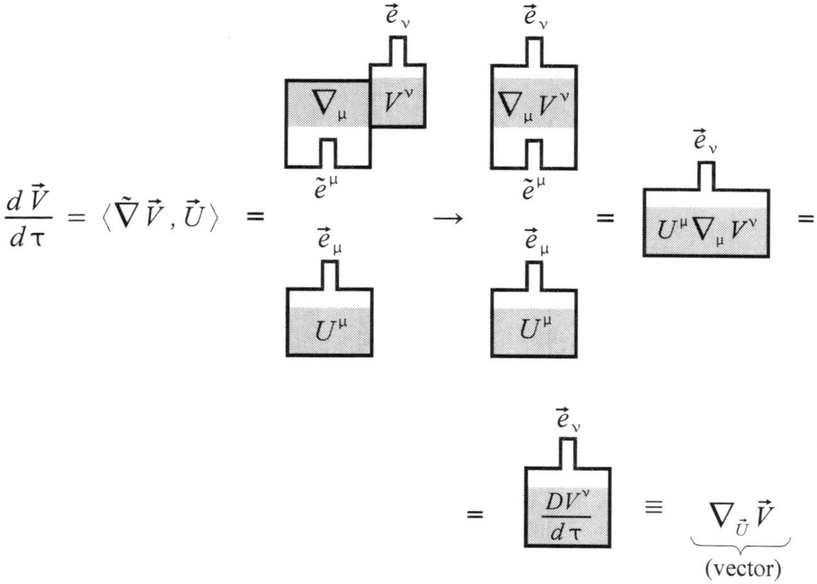

*Mnemo*

Derivative along a line of a vector: the previous block diagram can help recalling the right formulas.

In particular, writing $\nabla_{\vec{U}}\vec{V}$ recalls the disposition of the symbols in the block:

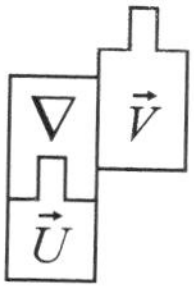


Also note that:

- $\dfrac{d\vec{V}}{d\tau} \equiv \langle \tilde{\nabla}\vec{V}, \vec{U} \rangle \equiv \nabla_{\vec{U}}\vec{V}$ contain the symbol → and are vectors
- $\dfrac{DV^{\nu}}{d\tau} \equiv U^{\mu}\nabla_{\mu}V^{\nu}$ are scalar components ( without symbol →)

*If a single thing is to keep in mind choose this:*

- directional covariant derivative $\equiv U^{\mu}\nabla_{\mu}$ : applies to both a scalar and vector components

## 5.4 Parallel transport along a line of a vector

Let $\vec{V}$ a vector defined in each point of a parametric line $x^{\mu}(\tau)$.

Moving from the point $P(\tau)$ to a point $P(\tau+\Delta\tau)$ the vector undergoes an increase:

$$\vec{V}(\tau+\Delta\tau) = \vec{V}(\tau) + \frac{d\vec{V}}{d\tau}(\tau)\Delta\tau + O[(\Delta\tau)^2] \; * \qquad 5.16$$

If the first degree term is missing, namely:

$$\vec{V}(\tau+\Delta\tau) = \vec{V}(\tau) + O[(\Delta\tau)^2] \quad \Leftrightarrow \quad \frac{d\vec{V}}{d\tau}=0 \qquad 5.17$$

we say that the vector $\vec{V}$ is ***parallel transported***.

Roughly: transporting its origin along the line for an amount $\Delta\tau$ the vector undergoes only a variation of the 2nd order (thus keeps itself almost unchanged).

The parallel transport of a vector $\vec{V}$ is defined locally, in its own flat neighborhood: over a finite length the parallelism between initial and transported vector is no longer maintained .

- In each point of the line let a tangent vector $\vec{U} = \dfrac{d\vec{x}}{d\tau}$ be defined.

The *parallel transport condition of* $\vec{V}$ *along* $\vec{U}$ can be expressed by

---

* We denote by $O(x)$ a quantity of the same order of $x$. In this case it is matter of a quantity of the same order of $(\Delta\tau)^2$, i.e. of the second order with respect to $\Delta\tau$.

one of the following forms:

$$\nabla_{\vec{U}} \vec{V} = 0 \quad , \quad \langle \tilde{\nabla} \vec{V}, \vec{U} \rangle = 0 \quad , \quad U^{\mu} \nabla_{\mu} V^{\nu} = 0 \qquad 5.18$$

all equivalent to the definition *eq.5.17* because (*eq.5.7*, *eq.5.13*):

$$\nabla_{\vec{U}} \vec{V} = \frac{d\vec{V}}{d\tau} = \langle \tilde{\nabla} \vec{V}, \vec{U} \rangle \overset{comp}{\rightarrow} U^{\mu} \nabla_{\mu} V^{\nu} \ .$$

The vector $\vec{V}$ to be transported can have any orientation with respect to $\vec{U}$. Given a vector $\vec{V}$ defined at a point of a regular line $x^{\mu}(\tau)$, it is always possible to parallel transport it along the line.

## 5.5 Geodesics

When it happens that the tangent vector $\vec{U}$ is transported parallel to itself along a line, the line is a ***geodesic***.

*Geodesic condition* for a line is then:

$$\frac{d\vec{U}}{d\tau} = 0 \quad \text{along the line} \qquad 5.19$$

Equivalent to *eq.5.19* are the following statements:

$$\nabla_{\vec{U}} \vec{U} = 0 \quad , \quad \langle \tilde{\nabla} \vec{U}, \vec{U} \rangle = 0 \qquad 5.20$$

or componentwise:

$$\left. \begin{aligned} & U^{\mu} \nabla_{\mu} U^{\nu} = 0 \\ & \frac{DU^{\nu}}{d\tau} = 0 \\ & \frac{dU^{\nu}}{d\tau} + \Gamma^{\nu}_{\mu\lambda} U^{\lambda} U^{\mu} = 0 \\ & \frac{d^2 x^{\nu}}{d\tau^2} + \Gamma^{\nu}_{\mu\lambda} \frac{dx^{\lambda}}{d\tau} \frac{dx^{\mu}}{d\tau} = 0 \end{aligned} \right\} \qquad 5.21$$

which come respectively from *eq.5.12*, *eq.5.7*, *eq.5.10*, *eq.5.11* (and *eq.5.1*). All *eq.5.19*, *eq.5.20*, *eq.5.21* are equivalent and represent the *equations of the geodesic*.

Some properties of geodesics are the following:

- Along a geodesic the tangent vector is constant in magnitude.*

---

* The converse is not true: the constancy in magnitude of the tangent vector is not a sufficient condition for a path to be a geodesic. See further on.

▫ Indeed: $\frac{d}{d\tau}(g_{\alpha\beta}U^\alpha U^\beta) = U^\mu \nabla_\mu (g_{\alpha\beta}U^\alpha U^\beta) =$

$$= U^\mu (g_{\alpha\beta}U^\alpha \nabla_\mu U^\beta + g_{\alpha\beta}U^\beta \nabla_\mu U^\alpha + U^\alpha U^\beta \nabla g_{\alpha\beta}) = 0$$

because $U^\mu \nabla_\mu U^\beta = U^\mu \nabla_\mu U^\alpha = 0$ is the parallel transport condition of $\vec{U}$ along the geodesic and $\nabla_\mu g_{\alpha\beta} = 0$ (*eq.4.67*).

Hence: $g_{\alpha\beta}U^\alpha U^\beta = const \Rightarrow \mathbf{g}(\vec{U},\vec{U}) \overset{\text{def}}{=} |\vec{U}|^2 = const$ ,q.e.d.

▪ A geodesic parameterized by a parameter $\tau$ is still a geodesic when re-parameterized * with an "*affine*" parameter $s$ like $s = a + b\tau$ (two parameters are called *affine* if linked by a linear relationship).

▪ The geodesics are the "straightest possible lines" in a curved space (see *Appendix*), and can be thought of as inertial trajectories of particles not subject to external forces; along them a particle transports its velocity vector parallel to itself, constant in magnitude and tangent to the path without acceleration.

In a generic curved space let be a geodesic parametrized with the time $t$ as parameter. The tangent vector is then the velocity vector: $\vec{U} = \frac{d\vec{x}}{dt} = \vec{v}$. Since the definition of geodesic $\frac{d\vec{v}}{dt} = 0$ , along the geodesic the acceleration is null:

$$\vec{a} = \frac{d\vec{v}}{dt} = 0 \quad \text{and} \quad a^\nu = \frac{Dv^\nu}{dt} = \frac{dv^\nu}{dt} + \Gamma^\nu_{\mu\lambda} v^\lambda v^\mu = 0 \qquad (eq.5.21)$$

(that happens because the two terms in $d/dt$ and $\Gamma$, in general $\neq 0$ and variable with the point, compensate, zeroing).
Hence geodesics are inertial paths, q.e.d.

We note *en passant* that the acceleration is the covariant derivative of the vector velocity $\vec{v}$ along the line.

We also note that the vector $\vec{v}$ is transported along the geodesics constant in magnitude: $|\vec{v}| = v = const$ . Hence, if $\tau$ is the arc-length, then $\tau = vt$ and the two parameters $\tau$ and $t$ are affine. Therefore taking the geodesic parametrized by $t$ is not limiting because the re-parameterization is still assured.

---

* A line may be parametrized in more than one way. For example, a parabola can be parametrized by $x = t;\ y = t^2$ or $x = t^3;\ y = t^6$.

A different description of the same motion is given thinking of the curved space as embedded inside a flat space (e.g. Euclidean) with an extra dimension: here the path is no longer a geodesic and the same motion is described as accelerated. In fact, in the new context is $\forall \Gamma = 0$ due to the flatness and therefore $a^{\nu} = \dfrac{d v^{\nu}}{d t} \neq 0$ and $\vec{a} \neq 0$.

The two different points of view suggest that the same motion can be described in terms of space curvature *or* acceleration (a typical example are planetary orbits). The equivalence between curvature and acceleration is one of the cornerstone of General Relativity.

▪ Along a geodesic the tangent vector (it may be, as seen, the velocity vector) is transported tangent to the curve and constant in magnitude. Conversely, it is worth to note that it is *not enough* for a curve to be geodesics that the tangent vector keeps tangent to it and constant in magnitude. For a curve to be geodesics it is required the parallel transport, which is a more stringent request.

For instance, on a 2D spherical surface geodesics are the greatest circles, i.e. the equator and the meridians, but not the parallels: along parallels the velocity vector can be transported tangent to the path and constant in magnitude, but the parallel transport does *not* occur.. As they are not geodetic paths, parallels are not inertial trajectories and can be traveled only if a constant acceleration directed along the meridian is in action.

▫ We work a concrete example in order to clarify a situation that is not immediately intuitive.
An aircraft flying along a parallel with speed $\vec{v}$ tangent to the parallel itself does *not* parallel-transport this vector along its path. A question arises about how would evolve a vector $\vec{E}$ at first coincident with $\vec{v}$ , but for which we require to be parallel transported all along the same trip.
For this purpose, as told about parallel transport, let's imagine to cut off from the spherical surface a narrow ribbon containing the path, in this case the whole Earth's parallel. This ribbon can be flattened and laid down on a plane; in this plane the path is an arc of a circumference (whose radius is the generatrix of the cone whose axis coincides with the Earth's axis and is tangent to

the spherical surface on the parallel); the length of the arc will be that of the Earth's parallel. Let's now parallel-transport on a plane the vector $\vec{E}_0$ initially tangent to the circumference until the end of the flattened arc (working on a plan, there is no ambiguity about the meaning of parallel transport). Let $\vec{E}_{fin}$ be the vector at the end of the transport. Now we imagine bringing back to its original place on the spherical surface the ribbon with the various subsequent positions taken by the vector $\vec{E}$ drawn on attached. In that way we achieve a graphical representation of the evolution of the vector $\vec{E}$ parallel transported along the Earth's parallel.

The steps of the procedure are explained in the figure below. After having traveled a full circle of parallel transport on the Earth's parallel of latitude $\alpha$ the vector $\vec{E}$ has undergone a clockwise rotation by an angle $\phi = \dfrac{2\pi r \cos\alpha}{r/\mathrm{tg}\,\alpha} = 2\pi \sin\alpha$ , from $\vec{E}_0$ to $\vec{E}_{fin}$ .

Only for $\alpha = 0$ (along the equator) we get $\phi = 0$ after a turn.

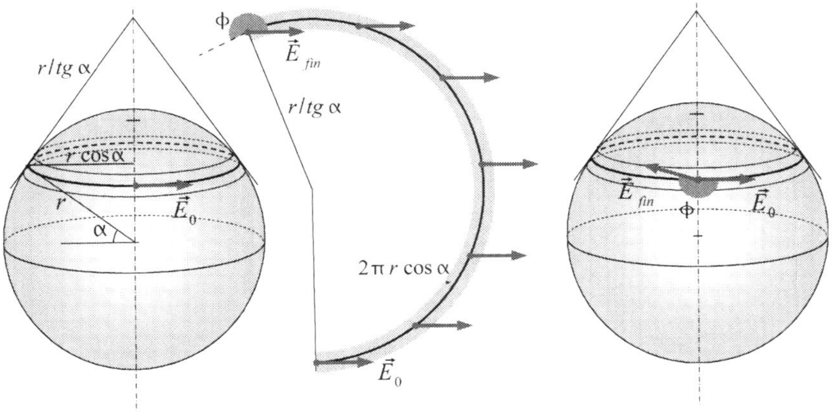


The trick to flatten on the plane the tape containing the trajectory works whatever is the initial orientation of the vector you want to parallel transport.

## 5.6 Positive and negative curvature

We have so far only considered the case of a "spherical surface" 2D space with constant curvature. There are, however, other types of curved spaces and other types of curvature. In fact, the curvature can be *positive* (e.g. a spherical space) or *negative* (hyperbolic space).
In a space with positive curvature the sum of the internal angles of a triangle is >180 °, the ratio of circumference to diameter is < π and two parallel lines end up with meeting each other; in a space with negative curvature the sum of the internal angles of a triangle is <180°, the ratio of circumference to diameter is > π and two parallel lines end up to diverge. In both cases Euclidean geometry is not valid.

▪ A space with negative curvature or hyperbolic curvature is less easily representable than a space with positive curvature or spherical. An intuitive approach is to imagine a thin flat metallic plate (a) that is unevenly heated. The thermal dilatation will be greater in the points where the higher is the temperature.

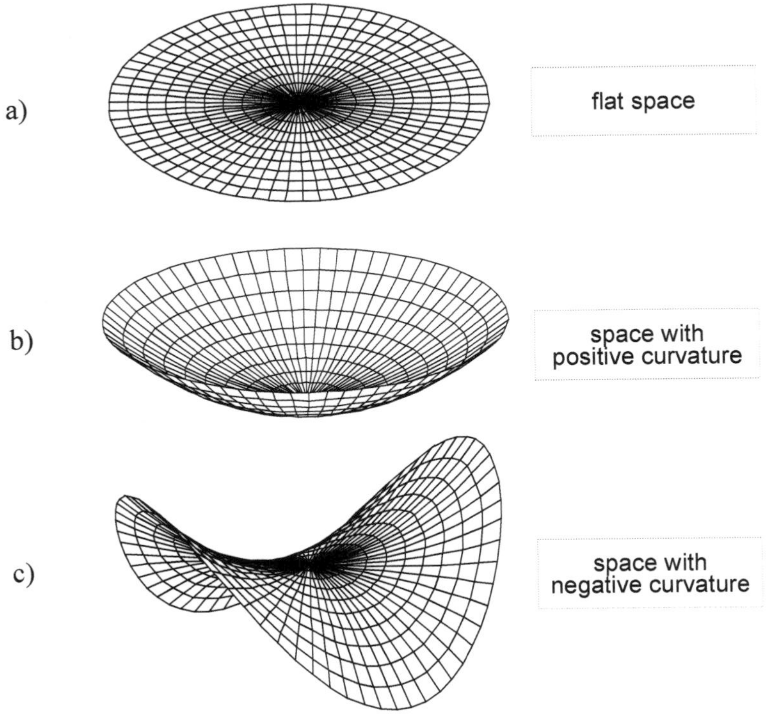

If the plate is heated more in the central part the termal dilating will be greater at the center and the foil will sag dome-shaped (b).
If the plate is more heated in the periphery, the termal dilation will be greater in the outer band and the plate will "embark" assuming a wavy shape (c).
The simplest case of negative curvature (c) is shown, in which the curved surface has the shape of a hyperbolic paraboloid (the form often taken by fried potato chips).

▪ For what concerns flat spaces, or zero-curvature manifolds, in addition to planes (or hyper-planes), other ones are part of the category. For example, among the 2D manifolds, the cylindrical surface and the conical surface are also flat; in general, are flat those 2D manifolds (surfaces) that can be laid without stretching on a plane (this is the case of the cylinder and the cone that can be developed in flat; it is *not* the case of a spherical surface). Generalization to any dimension is quite obvious even if no more intuitive. We note that also manifolds with a metric which is not positive definite can be flat, such as the Minkowski space of Special Relativity with a *diag* (± 1) metric.

## 5.7 Flat and curved manifold

The metric tensor **g** fully characterizes the geometry of the manifold, in particular its being flat or curved.* Now, either **g** is constant all over the manifold (i.e. it is unvarying) or varies with the point $P$. By definition:

$$\mathbf{g} \text{ constant} \quad \Leftrightarrow \quad \text{flat manifold} \qquad 5.22$$

and conversely:

$$\mathbf{g} = \mathbf{g}(P) \quad \Leftrightarrow \quad \text{curved manifold} \qquad 5.23$$

Since **g** is a tensor, it is independent of the coordinate system, but its representation by components $g_{\mu\nu}$ depends on it. Now:

$$\forall g_{\mu\nu} \text{ constant} \Rightarrow \mathbf{g} \text{ constant, but not viceversa,}$$

hence $$\forall g_{\mu\nu} \text{ constant} \Rightarrow \text{flat manifold, but not viceversa.}$$

---

* It will be seen later on apropos of Riemann's tensor.
Unless otherwise noted, here and below we consider *positive definite* metrics, i.e. metrics such that always $ds^2>0$ (for indefinite metrics *eq.5.22* applies only to the right ⇒ and *eq.5.23* only to the left ⇐).

Then, despite the manifold is flat, that is **g** is constant all over the manifold, it may be that in *some* coordinate system its components $g_{\mu\nu}$ are *P*-dependent;* on the contrary it must always be there some $g_{\mu\nu}$ *P*-dependent in *all* coordinate systems to be sure that **g** is *P*-dependent too, or $\mathbf{g}=\mathbf{g}(P)$, and then the manifold is curved.

A practical criterion to state the global flatness / not flatness of a manifold is:

$\exists$ reference where $\forall g_{\mu\nu}=const$ $\Rightarrow$ flat manifold; otherwise curved.

> What is discriminating is the existence of a reference system where all $g_{\mu\nu}$ are independent of the point $P$, that is to say that all elements of the matrix $[g_{\mu\nu}]$ are numerical constants. If such a system exists, the manifold is flat, otherwise it is curved.

- Note however that flatness or not – are *intrinsic features of the manifold*, independent of the reference system.

- In a flat manifold, in the reference system where $\forall g_{\mu\nu}=const$, it is consequently:

$$\forall \frac{\partial g_{\mu\nu}}{\partial x^{\alpha}}=0$$

everywhere, and so on for second derivatives and the subsequent ones. In a curved manifold not all derivatives of the metric are null; always null are the covariant derivatives, due to the property $\tilde{\nabla}\mathbf{g}=0$ (*eq.4.67*) which is valid in any Riemann' space. Then derivatives of the metric $g_{\mu\nu}$ are equal and opposite in sign to the terms in $\Gamma$ that appear in the (null) covariant derivative.

- In a flat manifold it is also $\forall\,\Gamma=0$ due to *eq.4.73*.
- Because of a theorem of matrix calculus, any invertible symmetric matrix like $[g_{\mu\nu}]$ can be diagonalized and carried into canonical form $[\eta_{\mu'\nu'}]=diag(\pm 1)$ by means of a matrix $\Lambda$ in the following way: **

$$\Lambda^{\mu}_{\mu'}\Lambda^{\nu}_{\nu'}g_{\mu\nu}=\eta_{\mu'\nu'} \qquad 5.24$$

---

* It is useful to keep in mind as a good example the polar coordinates ($\rho$, $\theta$), where $g_{\theta\theta}=1/\rho$ despite the flatness of the plane.

** In matrix terms: $\Lambda^{T} g\,\Lambda=\eta$

(we use the symbol $\eta$ for the metric in canonical form *diag* (±1), a diagonal matrix whit only +1 or −1 in the main diagonal * ).

For a flat manifold, being the metric independent of the point, the matrix $\Lambda$ holds *globally* for the whole manifold.

It is easy to recognize in $\Lambda$ the matrix that operates the changes of basis, which is related to a coordinate transformation (it is not the transformation but it comes down). By performing this coordinate transformation, the metric will be transformed (in coordinated bases) according to *eq.5.24*.

A flat space therefore admits a coordinate system in which the metric tensor assumes canonical form

$$[g_{\mu\nu}] = [\eta_{\mu\nu}] = diag(\pm 1)$$

due to the matrix $\Lambda$ related to the transformation.

This property can be assumed as a definition of flatness:

> Is flat a manifold where $g_{\mu\nu}$ assumes everywhere the canonical $diag(\pm 1)$ form in some (appropriate) coordinate system.

In particular, if $[g_{\mu\nu}] = diag(+1)$ the manifold is an Euclidean space; if $[g_{\mu\nu}] = diag(\pm 1)$ with a single −1 the manifold is a Minkowski space. The occurrences of +1 and −1 in the main diagonal, or their sum called "signature", is a characteristic of the given space: transforming the coordinates the +1 and −1 can exchange their places in the diagonal but do not change in number (*Sylvester's theorem*).

## 5.8 Flat local system

It is an intuitive notion that a curved manifold appears flat in the neighborhood of each point (excluding cusps or angular points): just take a small enough neighborhood and curvature will be imperceptible.

We can define a concept of "local flatness" by adapting to a local context the demands made for the global flatness of the manifold.

For a curved space to be locally flat in a point $P$, two conditions must be fulfilled in $P$:

---

* Also included here are indefinite metrics. The positive definite metrics can be diagonalized to the Euclidean-Cartesian canonical form diag (+1).

$$\forall \frac{\partial g_{\mu\nu}}{\partial x^{\alpha}}(P)=0 \quad , \quad [g_{\mu\nu}]=[\eta_{\mu\nu}]=diag(\pm 1) \text{ at } P \qquad 5.25$$

Thus, there must be the possibility to define in the neighborhood of $P$ a coordinate system such that in $P$ all first derivatives of $g_{\mu\nu}$ are null and $[g_{\mu\nu}]$ assumes canonical form.

▪ For curved manifolds, being $g_{\mu\nu}$ $P$-dependent, it does not exist a coordinate system capable to carry the metric to the canonical form *diag* (±1) with null derivatives in the whole manifold. That can be done only *punctually* by means of a coordinate transformation and its related matrix $\Lambda$. But that works well only in that point: the coordinate transformation that has $\Lambda$ as related matrix good for $P$ does not work for all the other points of the manifold, where the metric remains non-stationary and non-canonical.

▪ Asking for null first derivatives means to ask for $g_{\mu\nu}$ to be "stationary" in $P$ , so that its value remains almost unchanged in the neighborhood and holds, except for infinitesimals of higher order, even locally in the neighborhood of the point. * This is enough to ensure that in $P$ the metric is flat. The demand $g_{\mu\nu}$ stationary is the "weak" analogy of the demand $g_{\mu\nu}$ constant made for a flat manifold as a whole. The variability of the metric in the neighborhood of $P$ is only due to the second derivatives that, unlike what was in a flat space, cannot all be null.

▪ Of the two conditions *eq.5.25* crucial is the first; the second is accessory. We can proceed in two steps: first transform from the current reference to a new one in which the derivatives of all $g_{\mu\nu}$ are null (at this point $g_{\mu\nu}$ are stationary coefficients); then bring the new $g_{\mu\nu}$ into canonical form applying twice a matrix $\Lambda$ as in *eq.5.24.* By itself, $\Lambda$ expresses a *linear* coordinate transformation that, just because linear, retains the nullity of the derivatives of $g_{\mu\nu}$ produced by the first transformation (the derivatives of the new $g_{\mu\nu}$ are linear combinations of previous ones, all null).

▪ It happens that, given a curved manifold and some reference system, there are locally flat points, that is, points where the metric

* An analogy is given by a function $f(x)$ which is stationary at its maximum or minimum points. In those points it is $f'=0$ and for small displacements to the right or to the left $f(x)$ only varies for infinitesimal of higher order.

"spontaneously" fulfill the local flatness conditions. Which are these points depends on the reference system: if you change the reference, the points of "spontaneous" flatness change. Ultimately, local flatness, while being a property of each point in all Riemann spaces, is manifested as a *feature of the reference system.*

All this is formalized in the following theorem.

## 5.9 Local flatness theorem

Given a differentiable manifold and any point $P$ in it, it is possible to find a coordinate system that reduces the metric to the form:

$$g_{\mu\nu}(P') = \eta_{\mu\nu}(P) + O\left[(x^{\kappa} - x_0^{\kappa})^2\right] \qquad 5.26$$

where $P'(x^{\kappa})$ is any point of the neighborhood of $P(x_0^{\kappa})$. * The term $O[...]$ is for an infinitesimal of the second order with respect to the displacement from $P(x_0^{\kappa})$ to $P'(x^{\kappa})$, that we have improperly denoted for simplicity by $(x^{\kappa} - x_0^{\kappa})$.

The above equation is merely a Taylor series expansion around the point $P$, missing of the first order term $(x^{\kappa} - x_0^{\kappa}) \dfrac{\partial g_{\mu\nu}}{\partial x^{\kappa}}(P)$ because the first derivatives are supposed to be null in $P$.

We explicitly observe that the neighborhood of $P$ containing $P'$ belongs to the manifold, *not* to the tangent space. On the other hand $\eta_{\mu\nu}$ is the (global) flat metric of the tangent space in $P$.

Roughly, the theorem states that the space is locally flat in the neighborhood of each point and there it can be confused with the tangent space to less than a second-order infinitesimal.

The analogy is with the Earth's surface, that can be considered locally flat in the neighborhood of every point.

▫ The proof of the theorem is a semi-constructive one: it can be shown that there is a coordinate transformation whose related matrix makes the metric canonical in $P$ with all its first derivatives null; then we'll try to rebuild the coordinate transformation and its related matrix (or at least the initial terms of the series expansions of its inverse).

We report here a scheme of proof.

---

* See note on *eq.5.16.*

▫ Given a manifold with a coordinate system $x^{\alpha}$ and a point $P$ in it, let's transform to new coordinates $x^{\mu'}$ with the intention to bring the metric into canonical form in $P$ with zero first derivatives.
At any point $P'$ of the $P$-neighborhood the transformation gives:

$$g_{\mu'\nu'}(P') = \Lambda^{\alpha}_{\mu'}\Lambda^{\beta}_{\nu'}\, g_{\alpha\beta}(P') \qquad 5.27$$

(denoted $\Lambda^{\mu'}_{\alpha}$ the related matrix $\Lambda$ (see *eq.3.7*, *eq.3.8*); to transform $g_{\alpha\beta}$ we have to use twice its transposed inverse).
In the last formula both $g$ and $\Lambda$ are meant to be calculated in $P'$.
In a differentiable manifold, what happens in $P'$ can be approximated by what happens in $P$ by means of a Taylor series. We expand in Taylor series around $P$ both members of *eq.5.27*, first the left: *

$$g_{\mu'\nu'}(P') = g_{\mu'\nu'}(P) + (x^{\gamma'} - x_0^{\gamma'})\, g_{\mu'\nu',\gamma'}(P) + \\ + \frac{1}{2}(x^{\gamma'} - x_0^{\gamma'})(x^{\lambda'} - x_0^{\lambda'})\, g_{\mu'\nu',\gamma'\lambda'}(P) + \dots \qquad 5.28$$

and then the right one:

$$\Lambda^{\alpha}_{\mu'}\Lambda^{\beta}_{\nu'} g_{\alpha\beta}(P') = \Lambda^{\alpha}_{\mu'}\Lambda^{\beta}_{\nu'} g_{\alpha\beta}(P) + (x^{\gamma'} - x_0^{\gamma'})\frac{\partial}{\partial x^{\gamma'}}(\Lambda^{\alpha}_{\mu'}\Lambda^{\beta}_{\nu'} g_{\alpha\beta})(P) + \\ + \frac{1}{2}(x^{\gamma'} - x_0^{\gamma'})(x^{\lambda'} - x_0^{\lambda'})\frac{\partial^2}{\partial x^{\gamma'}\partial x^{\lambda'}}(\Lambda^{\alpha}_{\mu'}\Lambda^{\beta}_{\nu'} g_{\alpha\beta})(P) + \dots \qquad 5.29$$

Now let's equal order by order the right terms of *eq.5.28* and *eq.5.29* :

---

* Henceforth we will often use comma , to denote partial derivatives.
*Eq.5.28* expresses Taylor's formula for the function of multiple variables $g_{\mu'\nu'}$ (in the present case the $n$ variables $x^{\alpha}$).
For a function $f$ of 2 variables $x$, $y$ Taylor's formula is often written in the following form (denoting $f_{,x}$, $f_{,y}$, $f_{,xy}$, ... the partial derivatives):

$$f(x_0+h, y_0+k) = f(x_0, y_0) + f_{,x}(x_0, y_0)h + f_{,y}(x_0, y_0)k + \\ + \frac{1}{2}[f_{,xx}(x_0, y_0)h^2 + 2f_{,xy}(x_0, y_0)hk + f_{,yy}(x_0, y_0)k^2] + \dots$$

which is equivalent to:

$$f(P') = f(P) + (x - x_0)f_{,x}(P) + (y - y_0)f_{,y}(P) + \\ + \frac{1}{2}[(x - x_0)^2 f_{,xx} + (x - x_0)(y - y_0)f_{,xy} + (y - y_0)(x - x_0)f_{,yx} + (y - y_0)^2 f_{,yy}] + \dots$$

The latter form is suitable to be generalized to $n$ variables and, by adopting the sum convention for repeated indexes, leads directly to *eq.5.28*.

I) $g_{\mu'\nu'}(P) = \Lambda^{\alpha}_{\mu'}\Lambda^{\beta}_{\nu'}g_{\alpha\beta}(P)$ 5.30

II) $g_{\mu'\nu',\gamma'}(P) = \dfrac{\partial}{\partial x^{\gamma'}}(\Lambda^{\alpha}_{\mu'}\Lambda^{\beta}_{\nu'}g_{\alpha\beta})(P)$ 5.31

III) $g_{\mu'\nu',\gamma'\lambda'}(P) = \dfrac{\partial^2}{\partial x^{\gamma'}\partial x^{\lambda'}}(\Lambda^{\alpha}_{\mu'}\Lambda^{\beta}_{\nu'}g_{\alpha\beta})(P)$ 5.32

Hereafter all terms are calculated in $P$ and for that we omit this specification below.

Recalling that $\Lambda^{\alpha}_{\mu'} = \dfrac{\partial x^{\alpha}}{\partial x^{\mu'}}$ and carrying out the derivatives of product we get:

I) $g_{\mu'\nu'} = \dfrac{\partial x^{\alpha}}{\partial x^{\mu'}}\dfrac{\partial x^{\beta}}{\partial x^{\nu'}}g_{\alpha\beta}$ 5.33

II) $g_{\mu'\nu',\gamma'} = \dfrac{\partial^2 x^{\alpha}}{\partial x^{\gamma'}\partial x^{\mu'}}\dfrac{\partial x^{\beta}}{\partial x^{\nu'}}g_{\alpha\beta} + \dfrac{\partial x^{\alpha}}{\partial x^{\mu'}}\dfrac{\partial^2 x^{\beta}}{\partial x^{\gamma'}\partial x^{\nu'}}g_{\alpha\beta} + \dfrac{\partial x^{\alpha}}{\partial x^{\mu'}}\dfrac{\partial x^{\beta}}{\partial x^{\nu'}}g_{\alpha\beta,\gamma'}$ 5.34

III) $g_{\mu'\nu',\gamma'\lambda'} = \dfrac{\partial^3 x^{\alpha}}{\partial x^{\gamma'}\partial x^{\lambda'}\partial x^{\mu'}}\dfrac{\partial x^{\beta}}{\partial x^{\nu'}}g_{\alpha\beta} + \dots$ 5.35

The right side members of these equations contain terms such as $g_{\alpha\beta,\gamma'}$ that can be rewritten, using the chain rule, as $g_{\alpha\beta,\gamma'} = g_{\alpha\beta,\sigma}\dfrac{\partial x^{\sigma}}{\partial x^{\gamma'}}$ .

The typology of the right side member terms thus reduces to:

- known terms: the $g_{\alpha\beta}$ and their derivatives with respect to the old coordinates, such as $g_{\alpha\beta,\sigma}$
- unknowns to be determined: the derivatives of the old coordinates with respect to the new ones.

(We'll see that just the derivatives of various order of the old coordinates with respect to the new ones allow us to rebuild the matrix $\Lambda$ related to the transformation and the transformation itself.)

▫ Now let's perform the counting of equations and unknowns for the case $n = 4$ , the most interesting for General Relativity.
The equations are counted by the left side members of I, II, III ; the unknowns by the right side members of them.

I) There are 10 equations, as many as the independent elements of the $4 \times 4$ symmetric tensor $g_{\mu'\nu'}$. The $g_{\alpha\beta}$ are already known.

To get $g_{\mu'\nu'} = \eta_{\mu'\nu'} \Rightarrow$ we have to put $= 0$ or $\pm 1$ each of the 10 independent elements of $g_{\mu'\nu'}$ (in the right side member) so to get the desired metric $diag(\pm 1)$.

Consequently, among the 16 unknown elements of matrix $\Lambda$, 6 can be arbitrarily assigned, while the other 10 are determined by equations I (*eq.5.33*). (The presence of six degrees of freedom means that there is a multiplicity of transformations and hence of matrices $\Lambda$ able to make canonical the metric). In this way have been assigned or calculated values for all 16 first derivatives.

II) There are 40 equations ( $g_{\mu'\nu'}$ has 10 independent elements; $\gamma'$ can take 4 values). The independent second derivatives like $\dfrac{\partial^2 x^{\alpha}}{\partial x^{\gamma'} \partial x^{\mu'}}$ are 40 too (4 values for $\alpha$ at numerator; 10 different pairs of $\gamma'$, $\mu'$ at denominator).

All the $g_{\alpha\beta}$ and their first derivatives are already known. Hence, we can set $= 0$ all the 40 $g_{\mu'\nu',\gamma'}$ at first member (as it was in our intent) and determine all the 40 second derivatives as a consequence.

III) There are 100 equations (10 independent elements $g_{\mu'\nu'}$ to derive with respect to the 10 different pairs generated by 4 numbers). Known the other factors, 80 third derivatives like $\dfrac{\partial^3 x^{\alpha}}{\partial x^{\lambda'} \partial x^{\gamma'} \partial x^{\mu'}}$ are to be determined (the 3 indexes at denominator give 20 combinations, * the index at numerator 4 choices yet). Now, it is *not* possible to find a set of values for the 80 third derivatives such that all the 100 $g_{\mu'\nu',\gamma'\lambda'}$ vanish; 20 among them remain nonzero.

We have thus shown that, for any point $P$, by assigning appropriate values to the derivatives of various order, it is possible to get (more than) one metric $g_{\mu'\nu'}$ such that in this point:

- $g_{\mu'\nu'} = \eta_{\mu'\nu'}$, the metric of flat manifold $diag(\pm 1)$
- all its first derivatives are null: $\forall\, g_{\mu'\nu',\gamma'} = 0$
- some second derivatives are nonzero: $\exists$ some $g_{\mu'\nu',\gamma'\lambda'} \neq 0$

---

* It is a matter of combinations (in the case of 4 items) with repeats, different for at least one element.

▫ This metric $\eta_{\mu'\nu'}$ with zero first derivatives and second derivatives not all zero characterizes the flat local system in *P* (but the manifold itself is curved because of the nonzero second derivatives).

The metric $\eta_{\mu'\nu'}$ *with zero first and second derivatives* is instead the metric of the tangent space in *P* (which is a space everywhere flat). The flat local metric of the manifold and that of the tangent space coincide in *P* and (the metric being stationary) even in its neighborhood except for a difference of infinitesimals of higher order.

▫ Finally, we note that, once calculated or assigned appropriate values to the derivatives of various order of old coordinates with respect to the new ones, we can reconstruct the series expansions for the (inverse) transformation of coordinates $x^{\alpha}(x^{\mu'})$ and for the elements of matrix $\Lambda^{\alpha}_{\mu'}$ (transposed inverse of $\Lambda$). Indeed, in their series expansions in terms of new coordinates $x^{\mu'}$ around P :

$$x^{\alpha}(x^{\mu'}) \quad \rightarrow \quad x^{\alpha}(P') = x^{\alpha}(P) + (x^{\gamma'} - x_0^{\gamma'})\frac{\partial x^{\alpha}}{\partial x^{\gamma'}}(P) + \\ + \frac{1}{2}(x^{\gamma'} - x_0^{\gamma'})(x^{\lambda'} - x_0^{\lambda'})\frac{\partial^2 x^{\alpha}}{\partial x^{\gamma'}\partial x^{\lambda'}}(P) + \dots$$

$$\Lambda^{\alpha}_{\mu'}(x^{\mu'}) \quad \rightarrow \quad \Lambda^{\alpha}_{\mu'}(P') = \Lambda^{\alpha}_{\mu'}(P) + (x^{\gamma'} - x_0^{\gamma'})\frac{\partial \Lambda^{\alpha}_{\mu'}}{\partial x^{\gamma'}}(P) + \\ + \frac{1}{2}(x^{\gamma'} - x_0^{\gamma'})(x^{\lambda'} - x_0^{\lambda'})\frac{\partial^2 \Lambda^{\alpha}_{\mu'}}{\partial x^{\gamma'}\partial x^{\lambda'}}(P) + \dots = \\ = \frac{\partial x^{\alpha}}{\partial x^{\mu'}}(P) + (x^{\gamma'} - x_0^{\gamma'})\frac{\partial^2 x^{\alpha}}{\partial x^{\gamma'}\partial x^{\mu'}}(P) + \\ + \frac{1}{2}(x^{\gamma'} - x_0^{\gamma'})(x^{\lambda'} - x_0^{\lambda'})\frac{\partial^3 x^{\alpha}}{\partial x^{\gamma'}\partial x^{\lambda'}\partial x^{\mu'}}(P) + \dots$$

only the already known derivatives of the old coordinates with respect to the new ones appear as coefficients.

Once known its inverse, both the direct transformation $x^{\mu'}(x^{\alpha})$ and the related matrix $\Lambda = \left[\Lambda^{\mu'}_{\alpha}\right]$ that induce the canonical metric in *P* are in principle implicitly determined, as desired.

▪ It is worth noting that also in the flat local system the strategy "*comma goes to semicolon*" is applicable. Indeed, *eq.4.73* ensures that even in the flat local system it is $\forall \Gamma = 0$ and therefore covariant derivatives $\nabla_\mu$ and ordinary derivatives $\partial_\mu$ coincide.

- That enables us to get some results in a straightforward way.
  For example, in the flat local system in $P$ of a given manifold it is (like in all flat manifolds) $g_{\alpha\beta,\mu} = 0$; from $g_{\alpha\beta,\mu} = g_{\alpha\beta;\mu}$ it follows that $g_{\alpha\beta;\mu} \equiv \nabla_\mu g_{\alpha\beta} = 0$, which is the well-known result $\tilde{\nabla} \mathbf{g} = 0$. Since this equation is a tensorial one, if valid in the locally flat coordinate system, it is also valid in any other coordinate system for the same space.*
  In the following we will use again this strategy.

## 5.10 Riemann tensor

After parallel-transported a vector along a closed line in a curved manifold, the final vector differs from the initial by an amount $\Delta \vec{V}$ due to the curvature. This amount $\Delta \vec{V}$ depends on the path, but it can be used as a measure of the curvature in a point if calculated along a closed infinitesimal loop around that point.

Given a point $A$ of the $n$-dimensional manifold we build a "parallelogram loop" $ABCD$ leaning on the coordinate lines of two generic coordinates $x^\alpha, x^\beta$ picked among $n$ and we parallel-transport the vector $\vec{V}$ along the circuit $ABCDA_{fin}$**

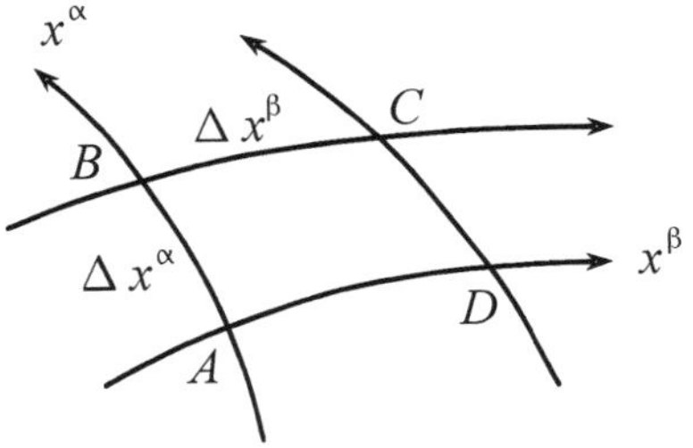


* In this particular case, since the manifold and its point $P$ are chosen arbitrarily, the equation has universal validity.

** The path $ABCDA_{fin}$ is a closed loop: in fact it lies on the surface or "hypersurface" in which the $n-2$ not involved coordinates are constant.

(of course, the construction should be thought as repeated for all possible pairs of coordinates with $\alpha, \beta = 1, 2, \ldots n$ ).

Tangent to the coordinates are the basis-vectors $\vec{e}_\alpha, \vec{e}_\beta$ , coordinated to $x^\alpha, x^\beta$ .

The parallel transport of $\vec{V}$ requires that $\dfrac{d\vec{V}}{d\tau} = 0$ or, component-wise $U^\alpha \nabla_\alpha V^\nu = 0$ *(eq.5.13)*, which along a segment of coordinate line $x^\alpha$ reduces to $\nabla_\alpha V^\nu = 0$ *(eq.5.15)*. But:

$$\nabla_\alpha V^\nu = 0 \quad \Rightarrow \quad \frac{\partial V^\nu}{\partial x^\alpha} + \Gamma^\nu_{\alpha\lambda} V^\lambda = 0 \quad \Rightarrow \quad \frac{\partial V^\nu}{\partial x^\alpha} = -\Gamma^\nu_{\alpha\lambda} V^\lambda \qquad 5.36$$

and a similar relation is obtained for the transport along $x^\beta$.

Due to the transport along the line segment *AB*, the components of the vector $\vec{V}$ undergo an increase:

$$V^\nu(B) = V^\nu(A) + \left(\frac{\partial V^\nu}{\partial x^\alpha}\right)_{(AB)} \Delta x^\alpha = V^\nu(A) - \left(\Gamma^\nu_{\alpha\lambda} V^\lambda\right)_{(AB)} \Delta x^\alpha \qquad 5.37$$

where *(AB)* means "calculated in an intermediate point between *A* and *B*". Along each of the four segments of the path there are variations:

$$\begin{aligned}
AB: &\qquad V^\nu(B) = V^\nu(A) - \left(\Gamma^\nu_{\alpha\lambda} V^\lambda\right)_{(AB)} \Delta x^\alpha \\
BC: &\qquad V^\nu(C) = V^\nu(B) - \left(\Gamma^\nu_{\beta\lambda} V^\lambda\right)_{(BC)} \Delta x^\beta \\
CD: &\qquad V^\nu(D) = V^\nu(C) + \left(\Gamma^\nu_{\alpha\lambda} V^\lambda\right)_{(CD)} \Delta x^\alpha \\
DA_{fin}: &\qquad V^\nu(A_{fin}) = V^\nu(D) + \left(\Gamma^\nu_{\beta\lambda} V^\lambda\right)_{(DA)} \Delta x^\beta
\end{aligned}$$

(in the last two segments $\Delta x^\alpha, \Delta x^\beta$ are negative in sign). Adding up member to member, terms $V^\nu(B)$, $V^\nu(C)$, $V^\nu(D)$ cancel out; hence:

$$\begin{aligned}
V^\nu(A_{fin}) - V^\nu(A) &= \left(\Gamma^\nu_{\beta\lambda} V^\lambda\right)_{(DA)} \Delta x^\beta - \left(\Gamma^\nu_{\beta\lambda} V^\lambda\right)_{(BC)} \Delta x^\beta + \\
&\quad + \left(\Gamma^\nu_{\alpha\lambda} V^\lambda\right)_{(CD)} \Delta x^\alpha - \left(\Gamma^\nu_{\alpha\lambda} V^\lambda\right)_{(AB)} \Delta x^\alpha \\
&= \left[\left(\Gamma^\nu_{\beta\lambda} V^\lambda\right)_{(DA)} - \left(\Gamma^\nu_{\beta\lambda} V^\lambda\right)_{(BC)}\right] \Delta x^\beta + \\
&\quad + \left[\left(\Gamma^\nu_{\alpha\lambda} V^\lambda\right)_{(CD)} - \left(\Gamma^\nu_{\alpha\lambda} V^\lambda\right)_{(AB)}\right] \Delta x^\alpha
\end{aligned}$$

using again the finite-increments theorem:

$$\left(\Gamma^{\nu}_{\beta\lambda}V^{\lambda}\right)_{(BC)} = \left(\Gamma^{\nu}_{\beta\lambda}V^{\lambda}\right)_{(DA)} + \frac{\partial}{\partial x^{\alpha}}\left(\Gamma^{\nu}_{\beta\lambda}V^{\lambda}\right)\Delta x^{\alpha}$$

$$\left(\Gamma^{\nu}_{\beta\lambda}V^{\lambda}\right)_{(AB)} = \left(\Gamma^{\nu}_{\beta\lambda}V^{\lambda}\right)_{(CD)} - \frac{\partial}{\partial x^{\alpha}}\left(\Gamma^{\nu}_{\beta\lambda}V^{\lambda}\right)\Delta x^{\beta}$$

where the derivatives are now calculated in intermediate points inside the loop, omitting to indicate it. We get:

$$= -\frac{\partial}{\partial x^{\alpha}}\left(\Gamma^{\nu}_{\beta\lambda}V^{\lambda}\right)\Delta x^{\alpha}\Delta x^{\beta} + \frac{\partial}{\partial x^{\beta}}\left(\Gamma^{\nu}_{\alpha\lambda}V^{\lambda}\right)\Delta x^{\beta}\Delta x^{\alpha}$$

$$= -\left(\Gamma^{\nu}_{\beta\lambda,\alpha}V^{\lambda} + \Gamma^{\nu}_{\beta\lambda}\frac{\partial V^{\lambda}}{\partial x^{\alpha}}\right)\Delta x^{\alpha}\Delta x^{\beta} + \left(\Gamma^{\nu}_{\alpha\lambda,\beta}V^{\lambda} + \Gamma^{\nu}_{\alpha\lambda}\frac{\partial V^{\lambda}}{\partial x^{\beta}}\right)\Delta x^{\beta}\Delta x^{\alpha}$$

and using the already known result $\frac{\partial V^{\lambda}}{\partial x^{\alpha}} = -\Gamma^{\lambda}_{\alpha\mu}V^{\mu}$ *(eq.5.36)*:

$$= -\left(\Gamma^{\nu}_{\beta\lambda,\alpha}V^{\lambda} - \Gamma^{\nu}_{\beta\lambda}\Gamma^{\lambda}_{\alpha\mu}V^{\mu}\right)\Delta x^{\alpha}\Delta x^{\beta} + \left(\Gamma^{\nu}_{\alpha\lambda,\beta}V^{\lambda} - \Gamma^{\nu}_{\alpha\lambda}\Gamma^{\lambda}_{\beta\mu}V^{\mu}\right)\Delta x^{\beta}\Delta x^{\alpha}$$

Writing $\mu$ as dummy index instead of $\lambda$ in the 1st and 3rd term:

$$= -\left(\Gamma^{\nu}_{\beta\mu,\alpha}V^{\mu} - \Gamma^{\nu}_{\beta\lambda}\Gamma^{\lambda}_{\alpha\mu}V^{\mu}\right)\Delta x^{\alpha}\Delta x^{\beta} + \left(\Gamma^{\nu}_{\alpha\mu,\beta}V^{\mu} - \Gamma^{\nu}_{\alpha\lambda}\Gamma^{\lambda}_{\beta\mu}V^{\mu}\right)\Delta x^{\beta}\Delta x^{\alpha}$$

$$= V^{\mu}\Delta x^{\alpha}\Delta x^{\beta}\left(-\Gamma^{\nu}_{\beta\mu,\alpha} + \Gamma^{\nu}_{\beta\lambda}\Gamma^{\lambda}_{\alpha\mu} + \Gamma^{\nu}_{\alpha\mu,\beta} - \Gamma^{\nu}_{\alpha\lambda}\Gamma^{\lambda}_{\beta\mu}\right)$$

We conclude:

$$\Delta V^{\nu} = V^{\mu}\Delta x^{\alpha}\Delta x^{\beta}\underbrace{\left(-\Gamma^{\nu}_{\beta\mu,\alpha} + \Gamma^{\nu}_{\alpha\mu,\beta} + \Gamma^{\nu}_{\beta\lambda}\Gamma^{\lambda}_{\alpha\mu} - \Gamma^{\nu}_{\alpha\lambda}\Gamma^{\lambda}_{\beta\mu}\right)}_{R^{\nu}_{\mu\beta\alpha}} \qquad 5.38$$

$R^{\nu}_{\mu\beta\alpha}$ must be a tensor because the other factors are tensors as well as $\Delta V^{\nu}$. Its rank is $\binom{1}{3}$ for the rank balancing: $\binom{1}{0} = \binom{1}{0}\binom{1}{0}\binom{1}{0}\cdot\binom{1}{3}$ .

The position made in *eq.5.38* is rewritten changing the name of the indexes $\alpha\leftrightarrow\nu$ , $\beta\leftrightarrow\mu$ for compliance with the commonly adopted notation::

$$R^{\alpha}_{\beta\mu\nu} \stackrel{\text{def}}{=} \left(-\Gamma^{\alpha}_{\mu\beta,\nu} + \Gamma^{\alpha}_{\nu\beta,\mu} + \Gamma^{\alpha}_{\mu\lambda}\Gamma^{\lambda}_{\nu\beta} - \Gamma^{\alpha}_{\nu\lambda}\Gamma^{\lambda}_{\mu\beta}\right) \qquad 5.39$$

*Eq.5.38*, rewritten with new indexes is:

$$\Delta V^{\alpha} = V^{\beta}\Delta x^{\nu}\Delta x^{\mu}R^{\alpha}_{\beta\mu\nu} \qquad 5.40$$

or, being $\tilde{P}$ any covector:

$$\tilde{P}(\Delta\vec{V}) = P_\alpha\, \Delta V^\alpha = P_\alpha V^\beta \Delta x^\nu \Delta x^\mu R^\alpha_{\beta\mu\nu} = \mathbf{R}(\tilde{P}, \vec{V}, \vec{\Delta x}, \vec{\Delta x}) \qquad 5.41$$

or else, multiplying *eq.5.40* by $\vec{e}_\alpha$ :

$$\Delta\vec{V} = \vec{e}_\alpha V^\beta \Delta x^\nu \Delta x^\mu R^\alpha_{\beta\mu\nu} \qquad 5.42$$

Both expressions can be used to interpret the conformation of **R**. *

- *Eq.5.40* lends itself to a rough intuitive interpretation of the Riemann tensor as "something like to the ratio between a relative variation of the vector parallel-transported along the loop and the area of the loop $\Delta x^\nu \Delta x^\mu$ " (the interpretation is roughly approximate: here it is not dealt exactly with a relative variation $\Delta V^\alpha / V^\alpha$ and $R^\alpha_{\beta\mu\nu}$ is not a simple proportionality coefficient, but a tensor).

▪ When defining **R** it is always tacitly supposed a passage to the limit $\Delta x^\alpha, \Delta x^\beta \rightarrow 0$ : **R** and its components are related to the point $P$ to which the infinitesimal loop degenerates and are punctually defined.

▪ $\mathbf{R} \equiv R^\nu_{\mu\beta\alpha}\, \vec{e}_\nu \tilde{e}^\mu \tilde{e}^\beta \tilde{e}^\alpha$ , the ***Riemann tensor,*** expresses the link between the change undergone by a parallel transported vector along an (infinitesimal) loop around a point $P$ in a curved space and the area of the loop, and doing so it gives a measure of the local curvature of the space; it contains the whole information about the curvature. In general it depends on the point: $\mathbf{R} = \mathbf{R}(P)$ .

---

* *Eq.5.42* means **R** applied to an incomplete input list, as represented below. If completing the list we get *eq.5.41*.
*Eq.5.40* is nothing but *eq.5.41* componentwise written.

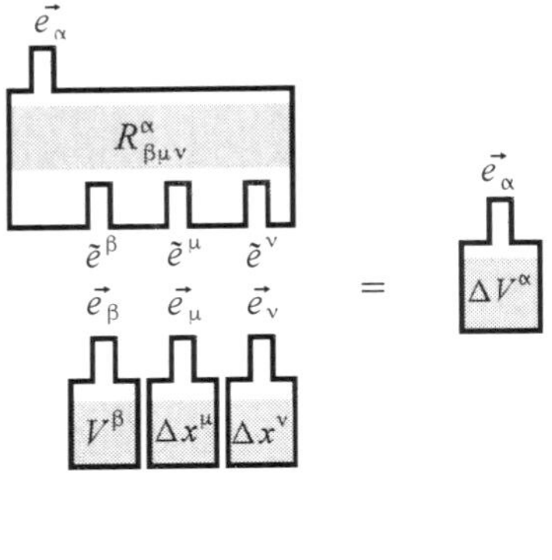

▪ If $\mathbf{R} = 0$ from *eq.5.41* it follows $\Delta \vec{V} = 0 \Rightarrow$ the vector $\vec{V}$ overlaps itself after the parallel transport along the infinitesimal loop around the point $P \Rightarrow$ the manifold is flat at that point. Hence:

$$\mathbf{R}(P) = 0 \quad \Rightarrow \quad \text{manifold flat in } P$$

If that holds globally the manifold is everywhere flat.
If $\mathbf{R} \neq 0$ in $P$, the manifold is curved at that point (although it is possible to define a local flat system in $P$).

▪ $\mathbf{R}$ *is a function of* $\Gamma$ *and its first derivatives* (*eq.5.39*), *namely of* $\mathbf{g}$ *and its first and second derivatives* (see *eq.4.73*).

▪ *In the flat local coordinate system* where $\mathbf{g}$ assumes the canonical form with zero first derivatives, $\mathbf{R}$ depends only on the second derivatives of $\mathbf{g}$. Let us specify now this dependence.
Because in any point of the manifold in its own flat local system it is $\forall \Gamma = 0$, the definition *eq.5.39* given for $R^{\alpha}_{\beta\mu\nu}$ reduces to:

$$R^{\alpha}_{\beta\mu\nu} = -\Gamma^{\alpha}_{\mu\beta,\nu} + \Gamma^{\alpha}_{\nu\beta,\mu} \qquad 5.43$$

From *eq.4.73* giving $\Gamma$ as functions of $g$ and its derivatives:

$\Gamma^{\alpha}_{\mu\beta} = \frac{1}{2} g^{\alpha\rho}(-g_{\mu\beta,\rho} + g_{\beta\rho,\mu} + g_{\rho\mu,\beta})$ and since $\frac{\partial}{\partial x^{\nu}} g^{\alpha\rho} = 0$ :

$$R^{\alpha}_{\beta\mu\nu} = -\frac{1}{2} g^{\alpha\rho}(-g_{\mu\beta,\rho\nu} + g_{\beta\rho,\mu\nu} + g_{\rho\mu,\beta\nu}) + $$
$$+\frac{1}{2} g^{\alpha\rho}(-g_{\nu\beta,\rho\mu} + g_{\beta\rho,\nu\mu} + g_{\rho\nu,\beta\mu})$$
$$= \frac{1}{2} g^{\alpha\rho}(+g_{\mu\beta,\rho\nu} - g_{\rho\mu,\beta\nu} - g_{\nu\beta,\rho\mu} + g_{\rho\nu,\beta\mu})$$

and multiplying both members by $g_{\alpha\kappa}$ we get (recall that $g_{\alpha\kappa} g^{\alpha\rho} = \delta^{\rho}_{\kappa}$ and hence $\rho \rightarrow \kappa$):

$$R_{\kappa\beta\mu\nu} = \frac{1}{2}(g_{\mu\beta,\kappa\nu} - g_{\kappa\mu,\beta\nu} - g_{\nu\beta,\kappa\mu} + g_{\kappa\nu,\beta\mu})$$

To write, as usual, $\alpha\,\beta$ as first pair and $\mu\,\nu$ as second it's enough to make the index exchange $\kappa \rightarrow \alpha$ ; by the symmetry of $g_{\alpha\beta}$ and rearranging the terms we get:

$$R_{\alpha\beta\mu\nu} = \frac{1}{2}(g_{\alpha\nu,\beta\mu} + g_{\beta\mu,\alpha\nu} - g_{\alpha\mu,\beta\nu} - g_{\beta\nu,\alpha\mu}) \qquad 5.44$$

that is true in each point of the manifold in its respective flat local system (i.e. when $g_{\alpha\beta}$ is the local flat metric), *not* in general).

▫ It is worth noting that the *eq.5.44* is no longer valid in a generic system because it is not a tensor equation *: outside the flat local system we need to retrieve the last two terms in $\Gamma\Gamma$ from *eq.5.39*. Therefore in a generic coordinate system it is:

$$R_{\alpha\beta\mu\nu}=\frac{1}{2}\left(g_{\alpha\nu,\beta\mu}+g_{\beta\mu,\alpha\nu}-g_{\alpha\mu,\beta\nu}-g_{\beta\nu,\alpha\mu}\right)+g_{\eta\alpha}\left(\Gamma^{\eta}_{\mu\lambda}\Gamma^{\lambda}_{\nu\beta}-\Gamma^{\eta}_{\nu\lambda}\Gamma^{\lambda}_{\mu\beta}\right) \qquad 5.45$$

*Mnemo*

To write down $R_{\alpha\beta\mu\nu}$ in the local flat system (*eq.5.44*) the "Pascal snail" can be used as a mnemonic aid to suggest the pairs of the indexes before the comma:

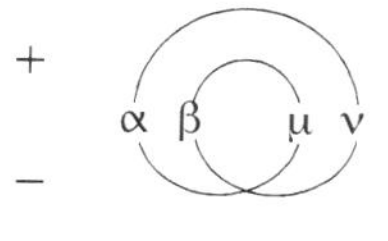


$+g_{\alpha\nu,..}+g_{\beta\mu,..}$

$-g_{\alpha\mu,..}-g_{\beta\nu,..}$

Use the remaining indexes for pairs after the comma.

▪ In $R_{\alpha\beta\mu\nu}$ it is usual to identify two pairs of indexes: $R_{\underbrace{\alpha\beta}_{1^{st}\text{ pair}}\underbrace{\mu\nu}_{2^{nd}\text{ pair}}}$

▪ According to common habit, also *eq.5.39* is reordered as:

$$R^{\alpha}_{\beta\mu\nu}=\Gamma^{\alpha}_{\beta\nu,\mu}-\Gamma^{\alpha}_{\beta\mu,\nu}+\Gamma^{\alpha}_{\lambda\mu}\Gamma^{\lambda}_{\beta\nu}-\Gamma^{\alpha}_{\lambda\nu}\Gamma^{\lambda}_{\beta\mu} \qquad 5.46$$

▪ The Riemann tensor **R** provides a flatness *criterion* more general than that one stated by *eq.5.22* e *eq.5.23* and also conclusive in cases of *P*-dependent metrics superimposed on flat spaces).

## 5.11 Symmetries of tensor R

The symmetries of $R_{\alpha\beta\mu\nu}$ can be studied in the flat local system by interchanging indexes in *eq.5.44*.

For example, exchanging the indexes $\alpha\leftrightarrow\beta$ within the first pair gives:

---

* Nor tensorial it can be made: $g_{\alpha\nu,\beta\mu}\neq g_{\alpha\nu;\beta\mu}$ even in the flat local system. Note that the trick "comma goes to semicolon" does not apply to second derivatives because in the double covariant derivative appear derivatives of the $\Gamma$ that are not null even in the flat local system where, $\partial_{\mu}\Gamma\neq 0$ despite $\Gamma=0$ .

$$R_{\beta\alpha\mu\nu} = \frac{1}{2}(g_{\beta\nu,\alpha\mu} + g_{\alpha\mu,\beta\nu} - g_{\beta\mu,\alpha\nu} - g_{\alpha\nu,\beta\mu})$$

whose right side member is the same as in the starting equation changed in sign and thus $\Rightarrow \quad R_{\beta\alpha\mu\nu} = -R_{\alpha\beta\mu\nu}$ .

It turns out from *eq.5.44* that **R** is: *i*) skew-symmetric with respect to the exchange of indexes within a pair; *ii*) symmetric with respect to the exchange of a pair with the other; it also enjoys a property such that: *iii*) the sum with cyclicity in the last 3 indexes is null:

*i*) $R_{\beta\alpha\mu\nu} = -R_{\alpha\beta\mu\nu} \qquad R_{\alpha\beta\nu\mu} = -R_{\alpha\beta\mu\nu}$ 5.47

*ii*) $R_{\mu\nu\alpha\beta} = R_{\alpha\beta\mu\nu}$ 5.48

*iii*) $R_{\alpha\beta\mu\nu} + R_{\alpha\mu\nu\beta} + R_{\alpha\nu\beta\mu} = 0$ 5.49

The four previous relations, although deduced in the flat local system (*eq.5.44* applies there only) are tensor equations and thus valid in any reference frame.

From *i*) it follows that the components with repeated indexes within the same pair are null (for example $R_{11\mu\nu} = R_{\beta\alpha 33} = R_{221\nu} = R_{1111} = 0$ ).
Indeed, from *i*): $R_{\alpha\alpha\mu\nu} = -R_{\alpha\alpha\mu\nu} \Rightarrow R_{\alpha\alpha\mu\nu} = 0$ ; and so on.

▪ On balance, because of its symmetries, among the $n^4$ components of $R_{\alpha\beta\mu\nu}$ only $n^2(n^2-1)/12$ are independent and $\neq 0$ (namely 1 for $n$ =2; 6 for $n$ = 3, 20 for $n$ = 4; ...).

## 5.12 Bianchi identity

It 's another relation linking the covariant first derivatives of $R_{\alpha\beta\mu\nu}$ :

$$R_{\alpha\beta\mu\nu;\lambda} + R_{\alpha\beta\nu\lambda;\mu} + R_{\alpha\beta\lambda\mu;\nu} = 0 \qquad 5.50$$

(the first two indexes α β are fixed; the others μ ν λ rotate).

▫ To get this result we place again in the flat local system of any point $P$ and calculate the derivatives of $R_{\alpha\beta\mu\nu}$ by *eq.5.44*:

$$R_{\alpha\beta\mu\nu,\lambda} = \frac{\partial}{\partial x^\lambda}\frac{1}{2}(g_{\alpha\nu,\beta\mu} + g_{\beta\mu,\alpha\nu} - g_{\alpha\mu,\beta\nu} - g_{\beta\nu,\alpha\mu})$$
$$= \frac{1}{2}(g_{\alpha\nu,\beta\mu\lambda} + g_{\beta\mu,\alpha\nu\lambda} - g_{\alpha\mu,\beta\nu\lambda} - g_{\beta\nu,\alpha\mu\lambda})$$

and similarly for $R_{\alpha\beta\nu\lambda,\mu}$ e $R_{\alpha\beta\lambda\mu,\nu}$ .

Adding up member to member the three equations and taking into account that in $g_{\alpha\beta,\gamma\delta\kappa}$ the indexes $\alpha\,\beta$ , $\gamma\,\delta$ * can be permuted at will within the two pairs, the right member vanishes so that:

$$R_{\alpha\beta\mu\nu,\lambda} + R_{\alpha\beta\nu\lambda,\mu} + R_{\alpha\beta\lambda\mu,\nu} = 0$$

Since in the flat local system there is no difference between (first) derivative and covariant derivative, namely:

$$R_{\alpha\beta\mu\nu;\lambda} = R_{\alpha\beta\mu\nu,\lambda} \;\;;\;\; R_{\alpha\beta\nu\lambda;\mu} = R_{\alpha\beta\nu\lambda,\mu} \;\;;\;\; R_{\alpha\beta\lambda\mu;\nu} = R_{\alpha\beta\lambda\mu,\nu}$$

we can write $R_{\alpha\beta\mu\nu;\lambda} + R_{\alpha\beta\nu\lambda;\mu} + R_{\alpha\beta\lambda\mu;\nu} = 0$ which is a tensor relationship and thus holds in any coordinate system, q.e.d.

## 5.12 Ricci tensor and Ricci scalar

$R_{\alpha\beta\mu\nu}$ may be contracted ** on indexes of the same pair or on indexes of different pairs. In the first case the result is 0, in the second it is significant.

These schemes illustrate some possible contractions and their results:

$$\left.\begin{array}{ccccc} R_{\alpha\beta\mu\nu} & \overset{\cdot g^{\alpha\beta}}{\longrightarrow} & R^{\beta}_{\beta\mu\nu} & \longrightarrow & R_{\mu\nu} \\ \downarrow = & & & & \\ -R_{\beta\alpha\mu\nu} & \overset{\cdot g^{\beta\alpha}}{\longrightarrow} & -R^{\alpha}_{\alpha\mu\nu} & \longrightarrow & -R_{\mu\nu} \end{array}\right\} \Rightarrow R_{\mu\nu} = 0$$

$$\left.\begin{array}{ccccc} R_{\alpha\beta\mu\nu} & \overset{\cdot g^{\alpha\mu}}{\longrightarrow} & R^{\mu}_{\beta\mu\nu} & \longrightarrow & R_{\beta\nu} \\ \downarrow = & & & & \\ R_{\mu\nu\alpha\beta} & \overset{\cdot g^{\mu\alpha}}{\longrightarrow} & R^{\alpha}_{\nu\alpha\beta} & \longrightarrow & R_{\nu\beta} \end{array}\right\} \Rightarrow R_{\beta\nu} = R_{\nu\beta} \qquad 5.51$$

Other similar cases still give as result 0 or $\pm R_{\beta\nu}$ .

We define ***Ricci tensor*** the $\binom{0}{2}$ tensor with components:

$R_{\beta\nu} \overset{\text{def}}{=}$ contraction of $R_{\alpha\beta\mu\nu}$ with respect to 1st and 3rd index 5.52

---

* $g_{\alpha\beta}$ is symmetric and the derivation order doesn't matter.

** In fact, it is $R^{\alpha}_{\beta\mu\nu}$ to undergo a contraction.

▪ Caution should be paid when contracting Riemann tensor on any two indexes: before contracting we must move the two indexes to contract in 1st and 3rd position using the symmetries of $R_{\alpha\beta\mu\nu}$, in order to perform anyway the contraction 1-3 whose result is known and positive by definition.
In such manner, due to the symmetries of $R_{\alpha\beta\mu\nu}$ and using schemes like *eq.5.51* we see that:

- have sign + the results of contraction on indexes 1-3 or 2-4
- have sign – the results of contraction on indexers 1-4 or 2-3

(while contractions on indexes 1-2 or 3-4 have 0 as result)

▪ The tensor $R_{\beta\nu}$ is symmetric (see *eq.5.51*).

▪ By further contraction we get the ***Ricci scalar*** $R$ :

$$R_{\beta\nu} \overset{\cdot g^{\beta\nu}}{\longrightarrow} R^{\nu}_{\nu} \longrightarrow R \quad (= \text{trace of } R_{\beta\nu}). \quad * \qquad 5.53$$

## 5.13 Einstein tensor

From Bianchi identity *eq.5.49*, raising and contracting twice the indexes to reduce Riemann to Ricci tensors and using the commutative property between index raising and covariant derivative:

$$g^{\beta\nu} g^{\alpha\mu} (R_{\alpha\beta\mu\nu;\lambda} + R_{\alpha\beta\nu\lambda;\mu} + R_{\alpha\beta\lambda\mu;\nu}) = 0$$

$$g^{\beta\nu} (R^{\mu}{}_{\beta\mu\nu;\lambda} + R^{\mu}{}_{\beta\nu\lambda;\mu} + R^{\mu}{}_{\beta\lambda\mu;\nu}) = 0$$

contracting $R^{\mu}{}_{\beta\mu\nu;\lambda}$ on indexes 1-3 (sign +) and $R^{\mu}{}_{\beta\lambda\mu;\nu}$ on indexes 1-4 (sign – ) :

$$g^{\beta\nu} (R_{\beta\nu;\lambda} + R^{\mu}{}_{\beta\nu\lambda;\mu} - R_{\beta\lambda;\nu}) = 0$$

using symmetry $R^{\mu}{}_{\beta\nu\lambda;\mu} = -R_{\beta}{}^{\mu}{}_{\nu\lambda;\mu}$ to allow the hooking and rising of index β :

$$g^{\beta\nu} (R_{\beta\nu;\lambda} - R_{\beta}{}^{\mu}{}_{\nu\lambda;\mu} - R_{\beta\lambda;\nu}) = 0$$

$$R^{\nu}{}_{\nu;\lambda} - R^{\nu\mu}{}_{\nu\lambda;\mu} - R^{\nu}_{\lambda;\nu} = 0$$

contracting $R^{\nu\mu}{}_{\nu\lambda;\mu}$ on indexes 1-3 (sign +)

---

* Remind that the trace is defined only for a mixed tensor $T^{\mu}_{\nu}$ as the sum $T^{\mu}_{\mu}$ of elements of the main diagonal of its matrix. Instead, the trace of $T_{\mu\nu}$ or $T^{\mu\nu}$ is by definition the trace of $g^{\kappa\mu} T_{\mu\nu}$ or, rispectively, of $g_{\kappa\mu} T^{\mu\nu}$.

$$R^{\nu}_{\nu;\lambda} - R^{\mu}_{\lambda;\mu} - R^{\nu}_{\lambda;\nu} = 0$$

The first term contracts again; the 2nd and 3rd term differ for a dummy index and are in fact the same:

$$R_{;\lambda} - 2\,R^{\mu}_{\lambda;\mu} = 0 \qquad 5.54$$

that, thanks to the identity $(2\,R^{\mu}_{\lambda} - \delta^{\mu}_{\lambda} R)_{;\mu} \equiv 2\,R^{\mu}_{\lambda;\mu} - R_{;\lambda}$ , may be written as:

$$(2\,R^{\mu}_{\lambda} - \delta^{\mu}_{\lambda} R)_{;\mu} = 0 \qquad 5.55$$

This expression or its equivalent *eq.5.54* are sometimes called *twice contracted Bianchi identities.*

Operating a further raising of the indexes:

$$g^{\nu\lambda}(2\,R^{\mu}_{\lambda} - \delta^{\mu}_{\lambda} R)_{;\mu} = 0$$

$$(2\,R^{\nu\mu} - \delta^{\nu\mu} R)_{;\mu} = 0$$

that, since $\delta^{\nu\mu} \equiv g^{\nu\mu}$ (*eq.2.46*), yields:

$$(2\,R^{\nu\mu} - g^{\nu\mu} R)_{;\mu} = 0$$

$$\left(R^{\nu\mu} - \frac{1}{2} g^{\nu\mu} R\right)_{;\mu} = 0 \qquad 5.56$$

We define ***Einstein tensor*** the $\binom{0}{2}$ tensor **G** whose components are: *

$$G^{\nu\mu} \stackrel{\text{def}}{=} R^{\nu\mu} - \frac{1}{2} g^{\nu\mu} R$$

The *eq.5.56* can then be written:

$$G^{\nu\mu}{}_{;\mu} = 0 \qquad 5.57$$

which shows that the tensor **G** has null divergence.

Since both $R^{\nu\mu}$ and $g^{\nu\mu}$ are symmetric under indexes exchange, **G** is symmetric too: $G^{\nu\mu} = G^{\mu\nu}$ .

The tensor **G** has an important role in General Relativity because it is the only divergence-free double-tensor that can be derived from **R** and as such contains the information about the curvature of the manifold.

---

* Nothing to do, of course, with the "dual switch" earlier denoted by the same symbol!

**G** *is our last stop. What's all that for? Blending physics and mathematics with an overdose of intuition, one day of a hundred years ago Einstein wrote:*

$$G^{\mu\nu} = \kappa \, T^{\mu\nu}$$

*anticipating, according to some, a theory of the Third Millennium to the twentieth century. This equation expresses a direct proportionality between the content of matter-energy represented by the "stress-energy-momentum" tensor* **T** *– a generalization in the Minkowski space-time of the stress tensor – and the curvature of space expressed by* **G** *(it was to ensure the conservation of energy and momentum that Einstein needed a zero divergence tensor like* **G***). In a sense,* **T** *belongs to the realm of physics, while* **G** *is matter of mathematics: that maths we have toyed with so far. Compatibility with the Newton's gravitation law allows to give a value to the constant* $\kappa$ *and write down the fundamental equation of General Relativity in the usual form:*

$$G^{\mu\nu} = \frac{8\pi G}{c^4} T^{\mu\nu}$$

*May be curious to note in the margin as, in this equation defining the fate of the Universe, beside fundamental constants such as the gravitational constant G and the light speed c, peeps out the ineffable* $\pi$*, which seems to claim in that manner its key role in the architecture of the world, although no one seems to make a great account of that (those who think of this observation as trivial try to imagine a Universe in which* $\pi$ *is, yes, a constant, but different from 3.14 ....).*

## Appendix

### 1 - *Transformation of* $\Gamma$ *under coordinate change*

We explicit $\Gamma^{\lambda}_{\mu\nu}$ from *eq.4.30* $\Gamma^{\lambda}_{\mu\nu}\vec{e}_{\lambda} = \partial_{\mu}\vec{e}_{\nu}$ by (scalar) multiplying both members by $\tilde{e}^{\kappa}$:

$$\Gamma^{\lambda}_{\mu\nu}\underbrace{\langle e_{\lambda}, \tilde{e}^{\kappa}\rangle}_{\delta^{\kappa}_{\lambda}} = \langle \partial_{\mu}\vec{e}_{\nu}, \tilde{e}^{\kappa}\rangle$$

$$\Gamma^{\kappa}_{\mu\nu} = \langle \partial_{\mu}\vec{e}_{\nu}, \tilde{e}^{\kappa}\rangle$$

Now let us operate a coordinate change $x^{\xi} \rightarrow x^{\zeta'}$ (and coordinate bases, too) and express the right member in the new frame:

$$\begin{aligned}
\Gamma^{\kappa}_{\mu\nu} &= \langle \partial_{\mu}\vec{e}_{\nu}, \tilde{e}^{\kappa}\rangle \\
&= \left\langle \frac{\partial x^{\alpha'}}{\partial x^{\mu}}\frac{\partial}{\partial x^{\alpha'}}(\Lambda^{\beta'}_{\nu}\vec{e}_{\beta'}),\ \Lambda^{\kappa}_{\gamma'}\tilde{e}^{\gamma'}\right\rangle \quad \text{(chain rule on } x^{\alpha'}\text{)} \\
&= \Lambda^{\alpha'}_{\mu}\left\langle \frac{\partial}{\partial x^{\alpha'}}(\Lambda^{\beta'}_{\nu}\vec{e}_{\beta'}),\ \Lambda^{\kappa}_{\gamma'}\tilde{e}^{\gamma'}\right\rangle \\
&= \Lambda^{\alpha'}_{\mu}\Lambda^{\kappa}_{\gamma'}\left\langle \frac{\partial}{\partial x^{\alpha'}}(\Lambda^{\beta'}_{\nu}\vec{e}_{\beta'}),\ \tilde{e}^{\gamma'}\right\rangle \\
&= \Lambda^{\alpha'}_{\mu}\Lambda^{\kappa}_{\gamma'}\left\langle \frac{\partial\vec{e}_{\beta'}}{\partial x^{\alpha'}}\Lambda^{\beta'}_{\nu} + \vec{e}_{\beta'}\frac{\partial}{\partial x^{\alpha'}}\Lambda^{\beta'}_{\nu},\ \tilde{e}^{\gamma'}\right\rangle \\
&= \Lambda^{\alpha'}_{\mu}\Lambda^{\kappa}_{\gamma'}\Lambda^{\beta'}_{\nu}\left\langle \frac{\partial\vec{e}_{\beta'}}{\partial x^{\alpha'}},\ \tilde{e}^{\gamma'}\right\rangle + \Lambda^{\alpha'}_{\mu}\Lambda^{\kappa}_{\gamma'}\frac{\partial}{\partial x^{\alpha'}}\Lambda^{\beta'}_{\nu}\langle \vec{e}_{\beta'}, \tilde{e}^{\gamma'}\rangle \\
&= \Lambda^{\alpha'}_{\mu}\Lambda^{\kappa}_{\gamma'}\Lambda^{\beta'}_{\nu}\underbrace{\left\langle \frac{\partial\vec{e}_{\beta'}}{\partial x^{\alpha'}},\ \tilde{e}^{\gamma'}\right\rangle}_{\Gamma^{\gamma'}_{\alpha'\beta'}} + \Lambda^{\alpha'}_{\mu}\Lambda^{\kappa}_{\gamma'}\frac{\partial}{\partial x^{\alpha'}}\Lambda^{\beta'}_{\nu}\delta^{\gamma'}_{\beta'} \\
&= \Lambda^{\alpha'}_{\mu}\Lambda^{\kappa}_{\gamma'}\Lambda^{\beta'}_{\nu}\Gamma^{\gamma'}_{\alpha'\beta'} + \Lambda^{\alpha'}_{\mu}\Lambda^{\kappa}_{\beta'}\frac{\partial}{\partial x^{\alpha'}}\Lambda^{\beta'}_{\nu} \\
&= \Lambda^{\alpha'}_{\mu}\Lambda^{\kappa}_{\gamma'}\Lambda^{\beta'}_{\nu}\Gamma^{\gamma'}_{\alpha'\beta'} + \frac{\partial x^{\alpha'}}{\partial x^{\mu}}\frac{\partial x^{\kappa}}{\partial x^{\beta'}}\frac{\partial^2 x^{\beta'}}{\partial x^{\alpha'}\partial x^{\nu}} \\
&= \Lambda^{\alpha'}_{\mu}\Lambda^{\kappa}_{\gamma'}\Lambda^{\beta'}_{\nu}\Gamma^{\gamma'}_{\alpha'\beta'} + \frac{\partial x^{\kappa}}{\partial x^{\beta'}}\frac{\partial^2 x^{\beta'}}{\partial x^{\mu}\partial x^{\nu}} \quad \text{(chain rule on } x^{\alpha'}\text{)}
\end{aligned}$$

The first term would describe a tensor transform, but the additional term leads to a different law and confirms that $\Gamma^{\kappa}_{\mu\nu}$ is not a tensor.

## *2 - Transformation of covariant derivative under coordinate change*

To $$V^{\kappa}_{;\mu} = V^{\kappa}_{,\mu} + \Gamma^{\kappa}_{\mu\nu} V^{\nu}$$

let's apply a coordinate change $x^{\xi} \rightarrow x^{\zeta'}$ and express the right member in the new coordinate frame:

$$V^{\kappa}_{;\mu} = V^{\kappa}_{,\mu} + \Gamma^{\kappa}_{\mu\nu} V^{\nu}$$

$$= \frac{\partial}{\partial x^{\mu}} (\Lambda^{\kappa}_{\gamma'} V^{\gamma'}) + (\text{transformation of } \Gamma)\, \Lambda^{\nu}_{\sigma'} V^{\sigma'}$$

$$= \Lambda^{\alpha'}_{\mu} \frac{\partial}{\partial x^{\alpha'}} (\Lambda^{\kappa}_{\gamma'} V^{\gamma'}) + (\text{transformation of } \Gamma)\, \Lambda^{\nu}_{\sigma'} V^{\sigma'}$$

$$= \Lambda^{\alpha'}_{\mu} \Lambda^{\kappa}_{\gamma'} \frac{\partial V^{\gamma'}}{\partial x^{\alpha'}} + V^{\gamma'} \Lambda^{\alpha'}_{\mu} \frac{\partial}{\partial x^{\alpha'}} (\Lambda^{\kappa}_{\gamma'}) +$$

$$+ \left( \Lambda^{\alpha'}_{\mu} \Lambda^{\kappa}_{\gamma'} \Lambda^{\beta'}_{\nu} \Gamma^{\gamma'}_{\alpha'\beta'} + \frac{\partial x^{\kappa}}{\partial x^{\beta'}} \frac{\partial^2 x^{\beta'}}{\partial x^{\mu} \partial x^{\nu}} \right) \Lambda^{\nu}_{\sigma'} V^{\sigma'}$$

$$= \Lambda^{\alpha'}_{\mu} \Lambda^{\kappa}_{\gamma'} \frac{\partial V^{\gamma'}}{\partial x^{\alpha'}} + V^{\gamma'} \Lambda^{\alpha'}_{\mu} \frac{\partial}{\partial x^{\alpha'}} (\Lambda^{\kappa}_{\gamma'}) +$$

$$+ \Lambda^{\alpha'}_{\mu} \Lambda^{\kappa}_{\gamma'} \Lambda^{\beta'}_{\nu} \Gamma^{\gamma'}_{\alpha'\beta'} \Lambda^{\nu}_{\sigma'} V^{\sigma'} + \frac{\partial x^{\kappa}}{\partial x^{\beta'}} \frac{\partial^2 x^{\beta'}}{\partial x^{\mu} \partial x^{\nu}} \Lambda^{\nu}_{\sigma'} V^{\sigma'}$$

$$= \Lambda^{\alpha'}_{\mu} \Lambda^{\kappa}_{\gamma'} \frac{\partial V^{\beta'}}{\partial x^{\alpha'}} + V^{\gamma'} \Lambda^{\alpha'}_{\mu} \frac{\partial}{\partial x^{\alpha'}} \frac{\partial x^{\kappa}}{\partial x^{\gamma'}} +$$

$$+ \Lambda^{\alpha'}_{\mu} \Lambda^{\kappa}_{\gamma'} \Gamma^{\gamma'}_{\alpha'\beta'} V^{\beta'} + \frac{\partial x^{\kappa}}{\partial x^{\beta'}} \frac{\partial^2 x^{\beta'}}{\partial x^{\mu} \partial x^{\sigma'}} V^{\sigma'} \quad *$$

$$= \Lambda^{\alpha'}_{\mu} \Lambda^{\kappa}_{\gamma'} \frac{\partial V^{\gamma'}}{\partial x^{\alpha'}} + V^{\gamma'} \frac{\partial x^{\alpha'}}{\partial x^{\mu}} \frac{\partial^2 x^{\kappa}}{\partial x^{\alpha'} \partial x^{\gamma'}} +$$

$$+ \Lambda^{\alpha'}_{\mu} \Lambda^{\kappa}_{\gamma'} \Gamma^{\gamma'}_{\alpha'\beta'} V^{\beta'} + \frac{\partial x^{\kappa}}{\partial x^{\beta'}} \frac{\partial^2 x^{\beta'}}{\partial x^{\mu} \partial x^{\sigma'}} V^{\sigma'}$$

or, changing some dummy indexes:

$$= \Lambda^{\alpha'}_{\mu} \Lambda^{\kappa}_{\gamma'} \frac{\partial V^{\gamma'}}{\partial x^{\alpha'}} + V^{\sigma'} \frac{\partial x^{\alpha'}}{\partial x^{\mu}} \frac{\partial^2 x^{\kappa}}{\partial x^{\alpha'} \partial x^{\sigma'}} +$$

$$+ \Lambda^{\alpha'}_{\mu} \Lambda^{\kappa}_{\gamma'} \Gamma^{\gamma'}_{\alpha'\beta'} V^{\beta'} + \frac{\partial x^{\kappa}}{\partial x^{\alpha'}} \frac{\partial^2 x^{\alpha'}}{\partial x^{\mu} \partial x^{\sigma'}} V^{\sigma'}$$

---

* Recall that $\Lambda^{\beta'}_{\nu} \Lambda^{\nu}_{\sigma'} = \frac{\partial x^{\beta'}}{\partial x^{\nu}} \frac{\partial x^{\nu}}{\partial x^{\sigma'}} = \frac{\partial x^{\beta'}}{\partial x^{\sigma'}} = \delta^{\beta'}_{\sigma'}$

$$= \Lambda^{\alpha'}_{\mu} \Lambda^{\kappa}_{\gamma'} V^{\gamma'}_{,\alpha'} + \Lambda^{\alpha'}_{\mu} \Lambda^{\kappa}_{\gamma'} \Gamma^{\gamma'}_{\alpha'\beta'} V^{\beta'} + \text{terms in } \partial^2$$

$$= \Lambda^{\alpha'}_{\mu} \Lambda^{\kappa}_{\gamma'} (V^{\gamma'}_{,\alpha'} + \Gamma^{\gamma'}_{\alpha'\beta'} V^{\beta'}) + \text{terms in } \partial^2$$

The two terms in $\partial^2$ cancel each other because opposite in sign. To show that let's compute:

$$\frac{\partial}{\partial x^{\sigma'}} \frac{\partial x^{\kappa}}{\partial x^{\mu}} = \frac{\partial}{\partial x^{\sigma'}} \left( \frac{\partial x^{\alpha'}}{\partial x^{\mu}} \frac{\partial x^{\kappa}}{\partial x^{\alpha'}} \right) = \frac{\partial x^{\alpha'}}{\partial x^{\mu}} \frac{\partial}{\partial x^{\sigma'}} \frac{\partial x^{\kappa}}{\partial x^{\alpha'}} + \frac{\partial x^{\kappa}}{\partial x^{\alpha'}} \frac{\partial}{\partial x^{\sigma'}} \frac{\partial x^{\alpha'}}{\partial x^{\mu}} =$$

$$= \frac{\partial x^{\alpha'}}{\partial x^{\mu}} \frac{\partial^2 x^{\kappa}}{\partial x^{\sigma'} \partial x^{\alpha'}} + \frac{\partial x^{\kappa}}{\partial x^{\alpha'}} \frac{\partial^2 x^{\alpha'}}{\partial x^{\sigma'} \partial x^{\mu}}$$

On the other hand: $\frac{\partial}{\partial x^{\sigma'}} \frac{\partial x^{\kappa}}{\partial x^{\mu}} = \frac{\partial}{\partial x^{\sigma'}} \delta^{\kappa}_{\mu} = 0$ , hence:

$$\frac{\partial x^{\alpha'}}{\partial x^{\mu}} \frac{\partial^2 x^{\kappa}}{\partial x^{\sigma'} \partial x^{\alpha'}} = -\frac{\partial x^{\kappa}}{\partial x^{\alpha'}} \frac{\partial^2 x^{\alpha'}}{\partial x^{\sigma'} \partial x^{\mu}} \text{ , q.e.d.}$$

The transformation for $V^{\kappa}_{;\mu}$ is then:

$$V^{\kappa}_{;\mu} = V^{\kappa}_{,\mu} + \Gamma^{\kappa}_{\mu\nu} V^{\nu} = \Lambda^{\alpha'}_{\mu} \Lambda^{\kappa}_{\gamma'} (V^{\gamma'}_{,\alpha'} + \Gamma^{\gamma'}_{\alpha'\beta'} V^{\beta'}) = \Lambda^{\alpha'}_{\mu} \Lambda^{\kappa}_{\gamma'} V^{\gamma'}_{;\alpha'} \Rightarrow$$

$\Rightarrow V^{\kappa}_{;\mu}$ transforms as a (component of a) $\binom{1}{1}$ tensor.

That allows us to define a tensor $\tilde{\nabla} \vec{V} \overset{comp}{\rightarrow} V^{\kappa}_{;\mu}$ .

### 3 - *Non-tensoriality of basis-vectors, their derivatives and gradients*

Are not tensors: a) basis-vectors $\vec{e}_\nu$ ; b) their derivatives $\partial_\mu \vec{e}_\nu$ ; c) their gradients $\tilde{\nabla}\vec{e}_\nu$

a) "Basis-vector" is a role that is given to certain vectors: within a vector space *n* vectors are chosen to wear the "jacket" of basis-vectors. Under change of basis these vectors remain unchanged, tensorially transforming their components (according to the usual contravariant scheme *eq.3.4*), while their role of basis-vectors is transferred to other vectors of the vector space. More than a law of transformation of basis-vectors, *eq.3.2* is the law ruling the role or jacket transfer.

For instance, under transformation from Cartesian coordinates to spherical, the vector $\vec{V} \equiv (1, 0, 0)$ that in Cartesian plays the role of basis-vector $\vec{i}$ transforms its components by *eq.3.4* and remains unchanged, but loses the role of basis-vector which is transferred to new vectors according to *eq.3.2* (which *looks like* a covariant scheme). Vectors underlying basis-vectors have then a tensorial character that *does not belong to basis-vectors as such.*

b) Scalar components of $\partial_\mu \vec{e}_\nu$ are the Christoffel symbols (*eq.4.30, eq.4.31*) which, as shown in Appendix 1, do not behave as tensors: that excludes that $\partial_\mu \vec{e}_\nu$ is a tensor.

The same conclusion is reached whereas the derivatives of basic-vectors $\partial_\mu \vec{e}_\nu$ are null in Cartesian coordinates but not in spherical ones. Since a tensor which is null in a reference must be zero in all, the *derivatives of basis-vectors* $\partial_\mu \vec{e}_\nu$ *are not tensors*.

c) Also the gradients of basis-vectors $\tilde{\nabla}\vec{e}_\nu$ have as scalar components the Christoffel symbols (*eq.4.51*); since they have not a tensorial character (see *Appendix 1*), it follows that *gradients of basis-vectors* $\tilde{\nabla}\vec{e}_\nu$ *are not tensors*.

As above, the same conclusion is reached whereas the $\tilde{\nabla}\vec{e}_\nu$ are zero in Cartesian but not zero in spherical coordinates.

We note that this does not conflict with the fact that $\partial\vec{V}, \tilde{\nabla}\vec{V}$ own tensor character: the scalar components of both are the covariant derivatives which transform as tensors.

Similar considerations apply to basis-covectors.

## 4 – *Equation of geodesic*

The equation of the geodesic is obtained as curve of minimal length between two points $A$ and $B$.

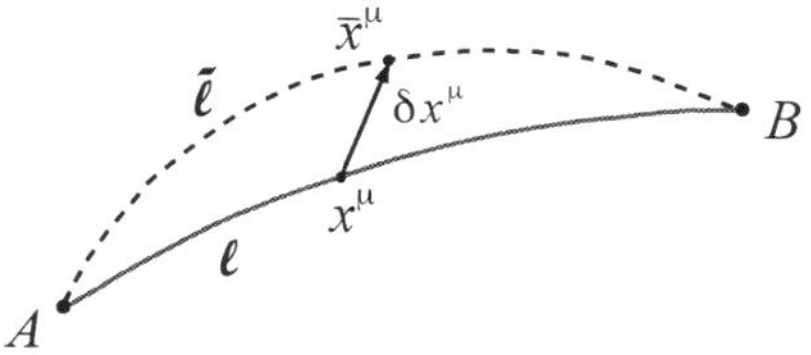


Let the geodesic $\ell$ between $A$ and $B$ be the parametric curve

$$x^\mu = x^\mu(t)$$

whose arc-lenght is

$$s = \int_{t_A}^{t_B} \sqrt{g_{\mu\nu} \frac{dx^\mu}{dt} \frac{dx^\nu}{dt}}\, dt$$

where in general $g_{\mu\nu}$ is function of the point, i.e. $g_{\mu\nu} = g_{\mu\nu}(x^\mu)$ .

Let's define in each point $x^\mu$ of the curve $\ell$ a small arbitrary vector $\vec{\delta x}$ variable with continuity, with components $\vec{\delta x} \overset{comp}{\rightarrow} \delta x^\mu$ that turns to zero in $A$ and $B$.

In this way a curve $\bar{\ell}$ is defined by equations

$$\bar{x}^\mu = x^\mu + \delta x^\mu \qquad \text{I}$$

whose arc-lenght is:

$$\bar{s} = \int_{t_A}^{t_B} \sqrt{\bar{g}_{\mu\nu} \frac{d\bar{x}^\mu}{dt} \frac{d\bar{x}^\nu}{dt}}\, dt$$

$\bar{g}_{\mu\nu}$ being now calculated in the points varied $\bar{x}^\mu$

$$\bar{g}_{\mu\nu} = \bar{g}_{\mu\nu}(\bar{x}^\mu)$$

The value of $\bar{g}_{\mu\nu}$ at point $\bar{x}^\mu$ can be referred to the value of $g_{\mu\nu}$ at point $x^\mu$ by series expansion (limited here to the first term):

$$\bar{g}_{\mu\nu} = g_{\mu\nu} + \frac{\partial g_{\mu\nu}}{\partial x^\xi} \delta x^\xi + \ldots \qquad (\xi = 1, 2, \ldots\, n)$$

The other two factors that appear under root can be written from *eq.I* as:

$$\frac{d\bar{x}^\mu}{dt} = \frac{d}{dt}(x^\mu + \delta x^\mu) = \frac{dx^\mu}{dt} + \frac{d(\delta x^\mu)}{dt}$$

$$\frac{d\bar{x}^{\nu}}{dt} = \frac{d}{dt}(x^{\nu} + \delta x^{\nu}) = \frac{dx^{\nu}}{dt} + \frac{d(\delta x^{\nu})}{dt}$$

The radicand is thus:

$$\bar{g}_{\mu\nu}\frac{d\bar{x}^{\mu}}{dt}\frac{d\bar{x}^{\nu}}{dt}\,dt = \left(g_{\mu\nu} + \frac{\partial g_{\mu\nu}}{\partial x^{\xi}}\,\delta x^{\xi}\right)\left(\frac{dx^{\mu}}{dt} + \frac{d(\delta x^{\mu})}{dt}\right)\left(\frac{dx^{\nu}}{dt} + \frac{d(\delta x^{\nu})}{dt}\right)$$

a product like $(A+a)(B+b)(C+c)$ where the capital means a finite term and the lowercase means an infinitesimal term. Carrying out the product:

$$(A+a)(B+b)(C+c) = \underline{ABC} + \underline{ABc} + \underline{AbC} + Abc + \underline{aBC} + aBc + abC + abc$$

Finite terms or first-order infinitesimal has been underlined; the others are infinitesimal of higher order, therefore negligible. So we get:

$$\bar{g}_{\mu\nu}\frac{d\bar{x}^{\mu}}{dt}\frac{d\bar{x}^{\nu}}{dt} = g_{\mu\nu}\frac{dx^{\mu}}{dt}\frac{dx^{\nu}}{dt} + g_{\mu\nu}\frac{dx^{\mu}}{dt}\frac{d(\delta x^{\nu})}{dt} + g_{\mu\nu}\frac{d(\delta x^{\mu})}{dt}\frac{dx^{\nu}}{dt} + \frac{\partial g_{\mu\nu}}{\partial x^{\xi}}\,\delta x^{\xi}\frac{dx^{\mu}}{dt}\frac{dx^{\nu}}{dt}$$

and being the 2nd and 3dt term the same one (just swap the dummies $\mu, \nu$ in the 3dt (here possible because $g_{\mu\nu}$ is simmetric):

$$\underbrace{\bar{g}_{\mu\nu}\frac{d\bar{x}^{\mu}}{dt}\frac{d\bar{x}^{\nu}}{dt}}_{a+\varepsilon} = \underbrace{g_{\mu\nu}\frac{dx^{\mu}}{dt}\frac{dx^{\nu}}{dt}}_{a,\ \text{preponderant term}} \underbrace{+2g_{\mu\nu}\frac{dx^{\mu}}{dt}\frac{d(\delta x^{\nu})}{dt} + \frac{\partial g_{\mu\nu}}{\partial x^{\xi}}\,\delta x^{\xi}\frac{dx^{\mu}}{dt}\frac{dx^{\nu}}{dt}}_{\varepsilon,\ \text{corrective term}}$$

From the calculation with small numbers it is known that: *

$$\sqrt{a+\varepsilon} \simeq \sqrt{a} + \frac{\varepsilon}{2\sqrt{a}} \quad \Rightarrow \quad \sqrt{a+\varepsilon} - \sqrt{a} \simeq \frac{\varepsilon}{2\sqrt{a}} \quad \text{thus:}$$

$$\sqrt{\bar{g}_{\mu\nu}\frac{d\bar{x}^{\mu}}{dt}\frac{d\bar{x}^{\nu}}{dt}} - \sqrt{g_{\mu\nu}\frac{dx^{\mu}}{dt}\frac{dx^{\nu}}{dt}} = \frac{g_{\mu\nu}\frac{dx^{\mu}}{dt}\frac{d(\delta x^{\nu})}{dt} + \frac{1}{2}\frac{\partial g_{\mu\nu}}{\partial x^{\xi}}\,\delta x^{\xi}\frac{dx^{\mu}}{dt}\frac{dx^{\nu}}{dt}}{\sqrt{g_{\mu\nu}\frac{dx^{\mu}}{dt}\frac{dx^{\nu}}{dt}}}$$

by integrating:

$$\delta s = \bar{s} - s = \int_{t_A}^{t_B}\sqrt{\bar{g}_{\mu\nu}\ldots}\,dt - \int_{t_A}^{t_B}\sqrt{g_{\mu\nu}\ldots}\,dt = \int_{t_A}^{t_B}\frac{g_{\mu\nu}\ldots}{\sqrt{g_{\mu\nu}\ldots}}\,dt$$

* First terms of Taylor expansion of $\sqrt{x}$ around $x = a$.

If the arc $s$ is chosen as parameter $t$ the denominator radicand is = 1 because $g_{\mu\nu}dx^{\mu}dx^{\nu}=ds^2$ and thus:

$$\delta s=\int_{s_A}^{s_B}\left(g_{\mu\nu}\frac{dx^{\mu}}{ds}\frac{d(\delta x^{\nu})}{ds}+\frac{1}{2}\frac{\partial g_{\mu\nu}}{\partial x^{\xi}}\delta x^{\xi}\frac{dx^{\mu}}{ds}\frac{dx^{\nu}}{ds}\right)ds=$$

$$=\int_{s_A}^{s_B}\underbrace{g_{\mu\nu}\frac{dx^{\mu}}{ds}}_{u}\underbrace{\frac{d(\delta x^{\nu})}{ds}ds}_{dv}+\int_{s_A}^{s_B}\frac{1}{2}\frac{\partial g_{\mu\nu}}{\partial x^{\xi}}\delta x^{\xi}\frac{dx^{\mu}}{ds}\frac{dx^{\nu}}{ds}ds=$$

Integration by parts $\int u\,dv=uv-\int v\,du$ yelds:

$$=\underbrace{\left[g_{\mu\nu}\frac{dx^{\mu}}{ds}\delta x^{\nu}\right]_{s_A}^{s_B}}_{=0}-\int_{s_A}^{s_B}\delta x^{\nu}\frac{d}{ds}\left(g_{\mu\nu}\frac{dx^{\mu}}{ds}\right)ds+\int_{s_A}^{s_B}\frac{1}{2}\frac{\partial g_{\mu\nu}}{\partial x^{\xi}}\delta x^{\xi}\frac{dx^{\mu}}{ds}\frac{dx^{\nu}}{ds}ds=$$

The term within brackets is è = 0 because $\delta x^{\nu}=0$ in $A$ and $B$; by swapping the dummies $\nu,\xi$ in the last integral:

$$=-\int_{s_A}^{s_B}\delta x^{\nu}\frac{d}{ds}\left(g_{\mu\nu}\frac{dx^{\mu}}{ds}\right)ds+\int_{s_A}^{s_B}\frac{1}{2}\frac{\partial g_{\mu\xi}}{\partial x^{\nu}}\delta x^{\nu}\frac{dx^{\mu}}{ds}\frac{dx^{\xi}}{ds}ds=$$

$$=-\int_{s_A}^{s_B}\delta x^{\nu}\left[\frac{d}{ds}\left(g_{\mu\nu}\frac{dx^{\mu}}{ds}\right)ds-\frac{1}{2}\frac{\partial g_{\mu\xi}}{\partial x^{\nu}}\frac{dx^{\mu}}{ds}\frac{dx^{\xi}}{ds}\right]ds$$

We calculate separately the term:

$$\frac{d}{ds}\left(g_{\mu\nu}\frac{dx^{\mu}}{ds}\right)=g_{\mu\nu}\frac{d^2x^{\mu}}{ds^2}+\frac{d\,g_{\mu\nu}}{ds}\frac{dx^{\mu}}{ds}=g_{\mu\nu}\frac{d^2x^{\mu}}{ds^2}+\frac{d\,g_{\mu\nu}}{dx^{\xi}}\frac{dx^{\mu}}{ds}\frac{dx^{\xi}}{ds}=$$

$$=g_{\mu\nu}\frac{d^2x^{\mu}}{ds^2}+\frac{1}{2}\frac{d\,g_{\mu\nu}}{dx^{\xi}}\frac{dx^{\mu}}{ds}\frac{dx^{\xi}}{ds}+\frac{1}{2}\frac{d\,g_{\xi\nu}}{dx^{\mu}}\frac{dx^{\xi}}{ds}\frac{dx^{\mu}}{ds}$$

having swapped the dummies $\mu,\xi$ in the last cut-in-half-term.
By replacing what calculated above:

$$\delta s=$$

$$=-\int_{s_A}^{s_B}\delta x^{\nu}\left[g_{\mu\nu}\frac{d^2x^{\mu}}{ds^2}+\frac{1}{2}\frac{d\,g_{\mu\nu}}{dx^{\xi}}\frac{dx^{\mu}}{ds}\frac{dx^{\xi}}{ds}+\frac{1}{2}\frac{d\,g_{\xi\nu}}{dx^{\mu}}\frac{dx^{\xi}}{ds}\frac{dx^{\mu}}{ds}-\frac{1}{2}\frac{\partial g_{\mu\xi}}{\partial x^{\nu}}\frac{dx^{\mu}}{ds}\frac{dx^{\xi}}{ds}\right]ds=$$

$$=-\int_{s_A}^{s_B}\delta x^{\nu}\left[g_{\mu\nu}\frac{d^2x^{\mu}}{ds^2}+\frac{dx^{\mu}}{ds}\frac{dx^{\xi}}{ds}\cdot\frac{1}{2}\left(g_{\mu\nu,\xi}+g_{\xi\nu,\mu}-g_{\mu\xi,\nu}\right)\right]ds$$

for $\delta s=\bar{s}-s=0$ (geodesic condition) the curve $\bar{\ell}$ overlaps the geodesic $\ell$ and the integral vanishes. Since $\delta x^{\mu}$ is arbitrary, to satisfy the geodesic condition $\delta s = 0$ the square bracket must vanish too:

$$g_{\mu\nu}\frac{d^2x^{\mu}}{ds^2}+\frac{dx^{\mu}}{ds}\frac{dx^{\xi}}{ds}\cdot\frac{1}{2}\left(g_{\mu\nu,\xi}+g_{\xi\nu,\mu}-g_{\mu\xi,\nu}\right)=0$$

Multiplying by $g^{\nu\lambda}$:

$$g_{\mu}^{\lambda}\frac{d^2x^{\mu}}{ds^2}+\frac{dx^{\mu}}{ds}\frac{dx^{\xi}}{ds}\cdot\underbrace{\frac{1}{2}g^{\nu\lambda}\left(g_{\mu\nu,\xi}+g_{\xi\nu,\mu}-g_{\mu\xi,\nu}\right)}_{\Gamma^{\lambda}_{\mu\xi}}=0$$

Since $g_{\mu}^{\lambda}=\delta_{\mu}^{\lambda} \quad\Rightarrow\quad g_{\mu}^{\lambda}\frac{d^2x^{\mu}}{ds^2}=\delta_{\mu}^{\lambda}\frac{d^2x^{\mu}}{ds^2}=\frac{d^2x^{\lambda}}{ds^2}$ we get the differential equation(s) of the geodesic:

$$\frac{d^2x^{\lambda}}{ds^2}+\frac{dx^{\mu}}{ds}\frac{dx^{\xi}}{ds}\Gamma^{\lambda}_{\mu\xi}=0$$

## 5 – *Riemannian metrics / spaces in general*

A space is not associated with a particular metric (the metric changes with the coordinate system you choose); however some characteristics of the metric can imply precise properties of space. This is the case of flat metrics (i.e. matrix with elements, or coefficients, constant) that imply flat space (but not viceversa) as well as defined positive / indefinite metrics that imply respectively space with positive distances ( $(ds^2 > 0)$ ) or with distances of variable sign.

Here we classify Riemannian spaces based on some properties of the spaces themselves or their metrics, referring to 4 attributes (0 = NO, 1 = YES). Forbidden combinations are barred.

| *g* positiv. defined | *g* const. coeffic. | flat space | Euclidean space | Examples or reasons for exclusion |
|---|---|---|---|---|
| 0 | 0 | 0 | 0 | example 1 |
| ~~0~~ | ~~0~~ | ~~0~~ | ~~1~~ | a |
| 0 | 0 | 1 | 0 | example 2 |
| ~~0~~ | ~~0~~ | ~~1~~ | ~~1~~ | b |
| ~~0~~ | ~~1~~ | ~~0~~ | ~~0~~ | c |
| ~~0~~ | ~~1~~ | ~~0~~ | ~~1~~ | a |
| 0 | 1 | 1 | 0 | Minkowski-like |
| ~~0~~ | ~~1~~ | ~~1~~ | ~~1~~ | b |
| 1 | 0 | 0 | 0 | example 3 |
| ~~1~~ | ~~0~~ | ~~0~~ | ~~1~~ | a |
| ~~1~~ | ~~0~~ | ~~1~~ | ~~0~~ | d |
| 1 | 0 | 1 | 1 | example 4 |
| ~~1~~ | ~~1~~ | ~~0~~ | ~~0~~ | c |
| ~~1~~ | ~~1~~ | ~~0~~ | ~~1~~ | a |
| ~~1~~ | ~~1~~ | ~~1~~ | ~~0~~ | d |
| 1 | 1 | 1 | 1 | Euclidean-rectangular |

("flat" and "Euclidean" refer here to space, *not* to the metric; "constant coefficients" means $\forall g_{\mu\nu} = const$; note that "constant coefficients" $\Leftrightarrow$ "flat metric" $\Rightarrow$ "flat space")

• Reason for exclusion:

a: Euclidean sp. $\Rightarrow$ flat sp.
b: Euclidean sp. $\Rightarrow$ $g$ positive definite
c: $g$ constant coefficients $\Rightarrow$ flat sp.
d: $\left.\begin{array}{l} g \text{ positive definite} \\ \text{flat sp.} \end{array}\right\} \Rightarrow$ Euclidean sp.

• Examples:

example 1: $[g_{\mu\nu}]=\begin{bmatrix}1 & 0\\ 0 & x^1\end{bmatrix} \qquad \Rightarrow \quad \exists\Gamma\neq 0\ ; \quad \mathbf{R}\neq 0$

example 2: $[g_{\mu\nu}]=\begin{bmatrix}1 & 0\\ 0 & -(x^2)^2\end{bmatrix} \qquad \Rightarrow \quad \exists\Gamma\neq 0\ ; \quad \mathbf{R}=0$

example 3: $[g_{\mu\nu}]=\begin{bmatrix}1 & 0\\ 0 & (x^1)^2+1\end{bmatrix} \qquad \Rightarrow \quad \exists\Gamma\neq 0\ ; \quad \mathbf{R}\neq 0$

example 4: $[g_{\mu\nu}]=\begin{bmatrix}1 & 0\\ 0 & (x^1)^2\end{bmatrix}$ (Euclidean metric in polar coordinates)
$\Rightarrow \quad \exists\Gamma\neq 0\ ; \quad \mathbf{R}=0$

▪ The whole set of cases has graphical representation in the diagram:

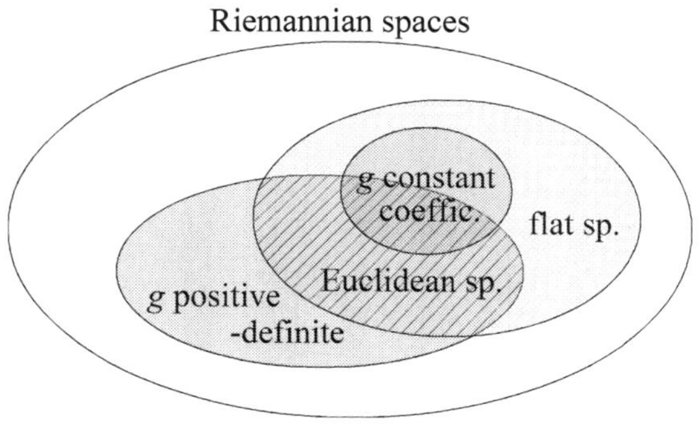


▪ For generic Riemannian metrics (undefined included) implications *eq.5.22* and *eq.5.23* change to:

$$\mathbf{g} \text{ constant} \Rightarrow \text{flat space}$$

$$\mathbf{g}=\mathbf{g}(P) \ \Leftarrow \text{curved space}$$

# Bibliographic references

- Among texts specifically devoted to Tensor Analysis the following retain a relatively soft profile:

Bertschinger, E. 1999, *Introduction to Tensor Calculus for General Relativity*, Massachusset Institute of Technology – Physics 8.962, pag. 34
Download: http://web.mit.edu/edbert/GR/gr1.pdf
In just over 30 pages a systematic, well structured and clear though concise treatment of Tensor Analysis aimed to General Relativity. The approach to tensors is of the geometric type, the best suited to acquire the basic concepts in a systematic way. The notation conforms to standard texts of Relativity. No use is made of matrix calculus. These notes are modeled in part on this text, possibly the best among those known to the author. Recommended without reserve; it can be at times challenging at first approach.

Kay, D. C. 1988, *Tensor Calculus*, McGraw-Hill, pag. 228
This book belongs to the mythical "Schaum Outline Series", but without having the qualities. The approach to Tensor Analysis is the most traditional "by components" (except a final chapter where the broad lines of the geometric approach are piled up in a very formal and almost incomprehensible summary). The traditional separation between statements and problems characteristic of Schaum's Outline is here interpreted in a not very happy way: the theoretical parts are somewhat abstract and formal, the problems are mostly examples of mechanisms not suitable to clarify the concepts. In fact, the conceptual part is the great absent in this book (which you can read almost to the end without having clear the very fundamental invariance properties of vectors and tensors subject to coordinate transformations!). Extensive use of matrix calculus (to which a far too insufficient review chapter is devoted). The second half of the book is dedicated to topics of interest for Differential Geometry but marginal for General Relativity. The notations used are not always standard. This text is not recommended as an introduction to the mathematics of General Relativity, to which it is rather misleading. Self-study risky. On the contrary it is no doubt useful for consultation of certain subjects, when the same are somewhat known.

Lovelock, D., Rund, H. 1989, *Tensors, Differential Forms and Variational Principles*, Dover Publication, pag. 366
Text with a somewhat mathematical setting, but still accessible and attentive to the conceptual propositions. The approach is that traditional "by components" ; are first introduced affine tensors and then general tensors according to a (questionable) policy of gradualism. The topics that are preliminary to General Relativity do not exceed one third of the book. It may be useful as a summary and reference for some single subjects.

Fleisch, D.A. 2012, *A Student's Guide to Vectors and Tensors*, Cambridge University Press, pag. 133
A very "friendly" introduction to Vector and Tensor Analysis, understandable even without special prerequisites, neverthless with a good completeness (until the introduction of the Riemann tensor, but without going into curved spaces). The approach to tensor is traditional. Beautiful illustrations, many examples from physics, many calculations carried out in full, together with a speech that gives the impression of proceeding methodically and safely, without jumps and without leaving behind dark spots, make this book an excellent autodidactict tool, also accessible to a good high school student.

Spain, B. 2003, *Tensor Calculus -A Concise Course*, Dover Publications, pag. 125
A dense booklet, not so easy and quite old-style, but comprehensive and interesting for the clever treatment of some topics (included the deduction of the equation of the geodesic without resorting Euler-Lagrange equation, we referred to in *Appendix 4*).

- Among the texts of General Relativity (GR), the following contains a sufficiently practicable introduction to Tensor Analysis carried on by a geometrical approach:

Schutz, B. F. 1985, *A First Course in General Relativity*, Cambridge University Press, pag. 376
Classic introductory text to General Relativity, it contains a quite systematic and complete discussion, although fragmented, of tensors. At least in the first part of the book (the one that concerns us here), the discourse is carefully argued, even meticulous, with all the explanations of the case and only few occasional disconnection (even if the impression is sometimes that of a bit cumbersome "machinery"). This is a good introduction to the topic which has the only defect of "flattening" significantly concepts and results without giving a different emphasis depending on their importance. For that it would be better dealing with it with some landmark already acquired. It has been used in various circumstances as a reference while writing these notes.

Dunsby, P. K. S. 2000 *An Introduction to Tensors and Relativity*, University of Cape Town, South Africa, pag. 130
Download: http://www.mth.uct.ac.za/omei/gr/ (PS format)
This is a schematic but well-made summary that follows the text of Schutz. With few variations, some simplifications and much more incisiveness.

Carroll, S.M. 2004 *Spacetime and geometry*, Addison-Wesley, pag. 513
Text that has established itself as a leader among those recently published on the subject GR. It's a fascinating example of scientific prose of conversational style: reading it sequentially you may have the impression of participating in a course of lectures. Tensors are introduced and developed closely integrated with the GR. This text places itself at a level of difficulty somewhat higher than that of Schutz.

Carroll, S. M. 1997 *Lecture Notes on General Relativity*, University of California Santa Barbara, pag. 231
Download: http://xxx.lanl.gov/PS_cache/gr-qc/pdf/9712/9712019v1.pdf
These are the original readings, in a even more conversational tone, from which the text quoted above has been developed. All that is important is located here, too, except for some advanced topics of GR.

Dettmann, C. P. 2007 *General Relativity*, Universty of Bristol, pag.36
Download: http://www.maths.bris.ac.uk/~macpd/gen_rel/bnotes.pdf
A thorough and original synthesis of the entire GR in few pages, including tensors. Readable, despite the compression. The structure of the text makes it easy to consult, even with regard to tensors.

McMahon, D. 2006 *Relativity Demystified*, Mc Graw-Hill, pag. 344
One of the "Self-teaching Guide"of the "Demystified" Series. It aims to achieve a "working knowledge" of the subject, without dwelling too much upon theory and subtilizing concepts. Pragmatic approach, sometimes a little rough, but interesting for a number of exercises developed in full. Useful, but not sufficient for those not satisfied by the approximative.

Ta-Pei Cheng 2010 *Relativity, Gravitation and Cosmology – A Basic introduction*, Oxford University Press, pag.435
A beautiful volume from the Oxford Master Series that deals exhaustively both GR and cosmology. The setting is teaching, and various topics are covered thoroughly, widely, and sometimes in a somewhat insisted way. In short, it is difficult not to understand, but first there is a lot to read. Tensors are first introduced in outline in the early chapters, but then, prior to Part IV "Relativity: full tensor formulation", two chapters are dedicated to them, albeit shrinking to what is interesting for GR. Appreciable discussion and examples of curved spaces. The book is also worthy of consideration for the considerable number of problems proposed, and in many cases solved.